THIS BOOK BELONGS TO

_____

_____

_____

Copyright © 2023 by SCH Alyssa, Siu Chin Hung and SCH Ali Good Fortune International Co

All rights reserved. No part of this publication may be reproduced, distributed, or transmitted, in any form or by any means, including photocopying, recording, or other electronic or mechanical methods, without prior written permission of the publisher, except in the case of brief quotations embodied in critical reviews and certain other noncommercial uses permitted by copyright law.

Rules of the game

The only thing you have to do is decrypt the secret meaningful message on every page.

In the beginning, we give the puzzle player some hints to start with, such as: Vowels

Most common takes the character that occurs the most and gives that one or some for free. Then, to increase the difficulty, the provided free number of characters will become lesser.

In addition, please note that not all alphabets will appear in the messages. Therefore, puzzle players will find it quite challenging. But I'm sure you'll be able to do it!

Check the answer found on the back of each quiz !

Discover how this puzzle type looks in practice by playing the below sample puzzle - an easy one for demonstration ●

Check the answer found on the back !

Answer.

ALYSSA IS MY DEAR DAUGHTER. I LOVE HER SO MUCH.

Quiz 1. What is this secret message about?
Difficulty: * * *

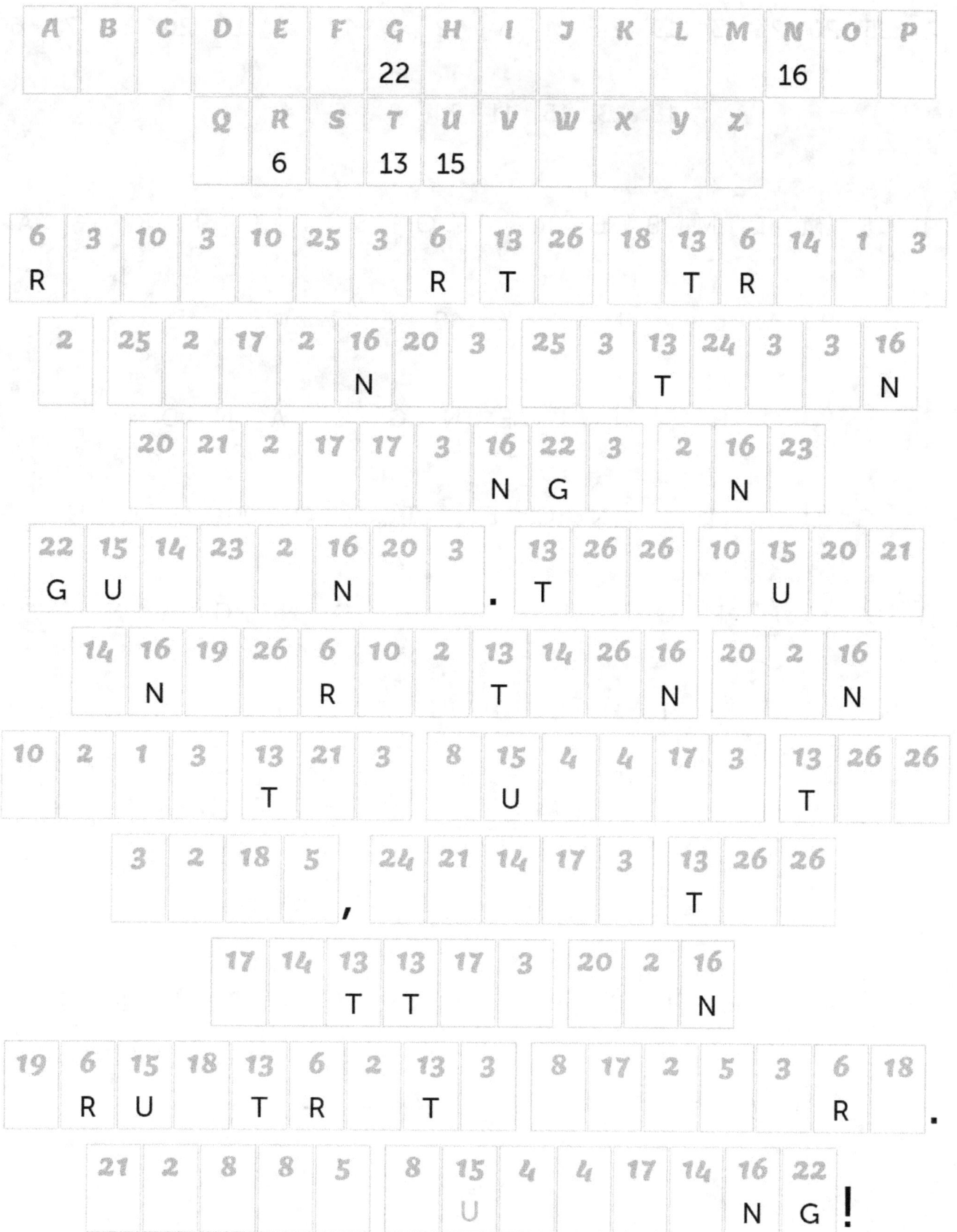

Answer.

REMEMBER TO STRIKE A BALANCE BETWEEN CHALLENGE AND GUIDANCE. TOO MUCH INFORMATION CAN MAKE THE PUZZLE TOO EASY, WHILE TOO LITTLE CAN FRUSTRATE PLAYERS. HAPPY PUZZLING!

Quiz 2. What is this secret message about?
Difficulty: * * *

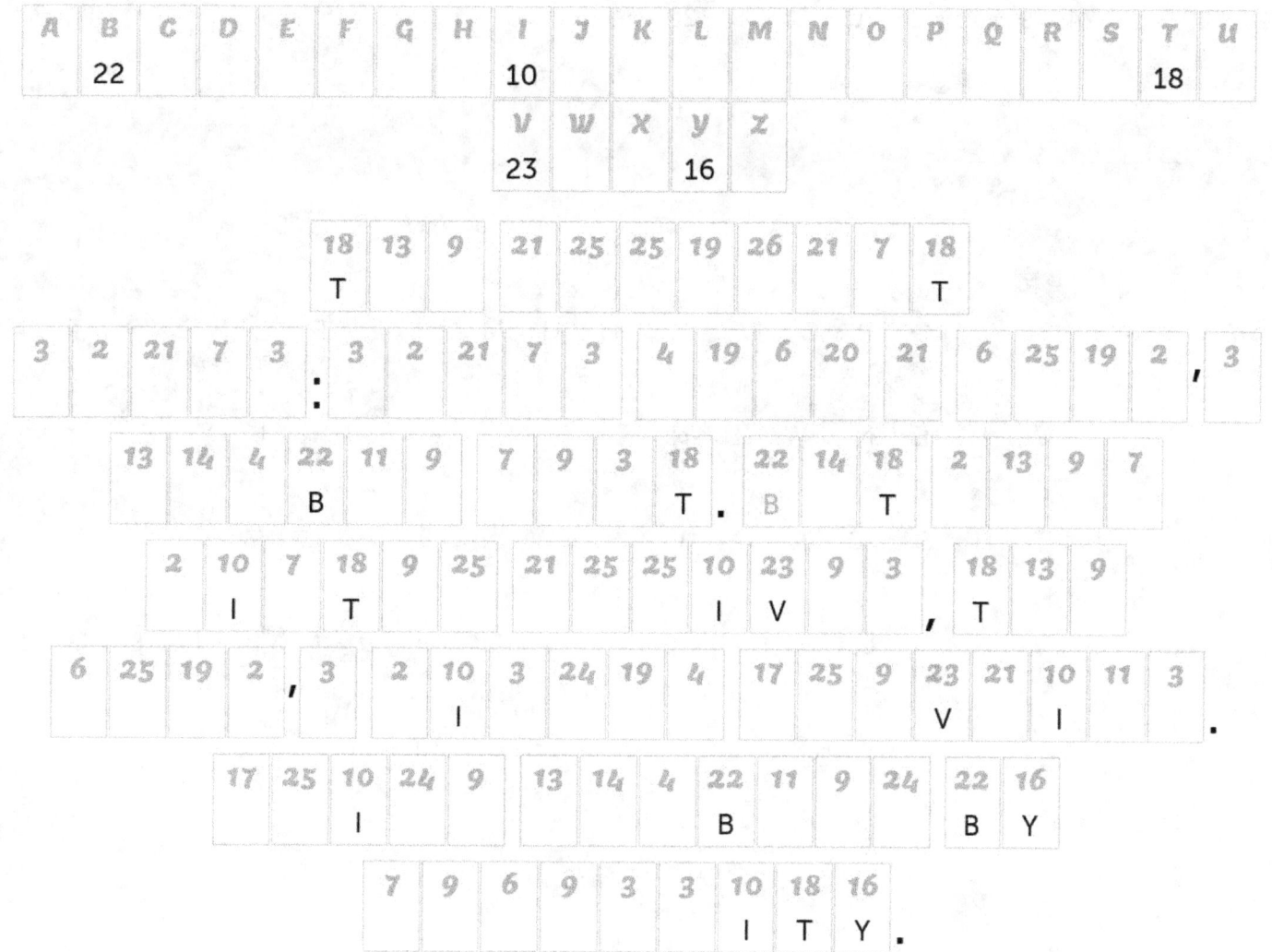

Answer.

THE ARROGANT SWANS: SWANS
MOCK A CROW'S HUMBLE NEST.
BUT WHEN WINTER ARRIVES,
THE CROW'S WISDOM
PREVAILS. PRIDE HUMBLED
BY NECESSITY.

Quiz 3. What is this secret message about?
Difficulty: * * *

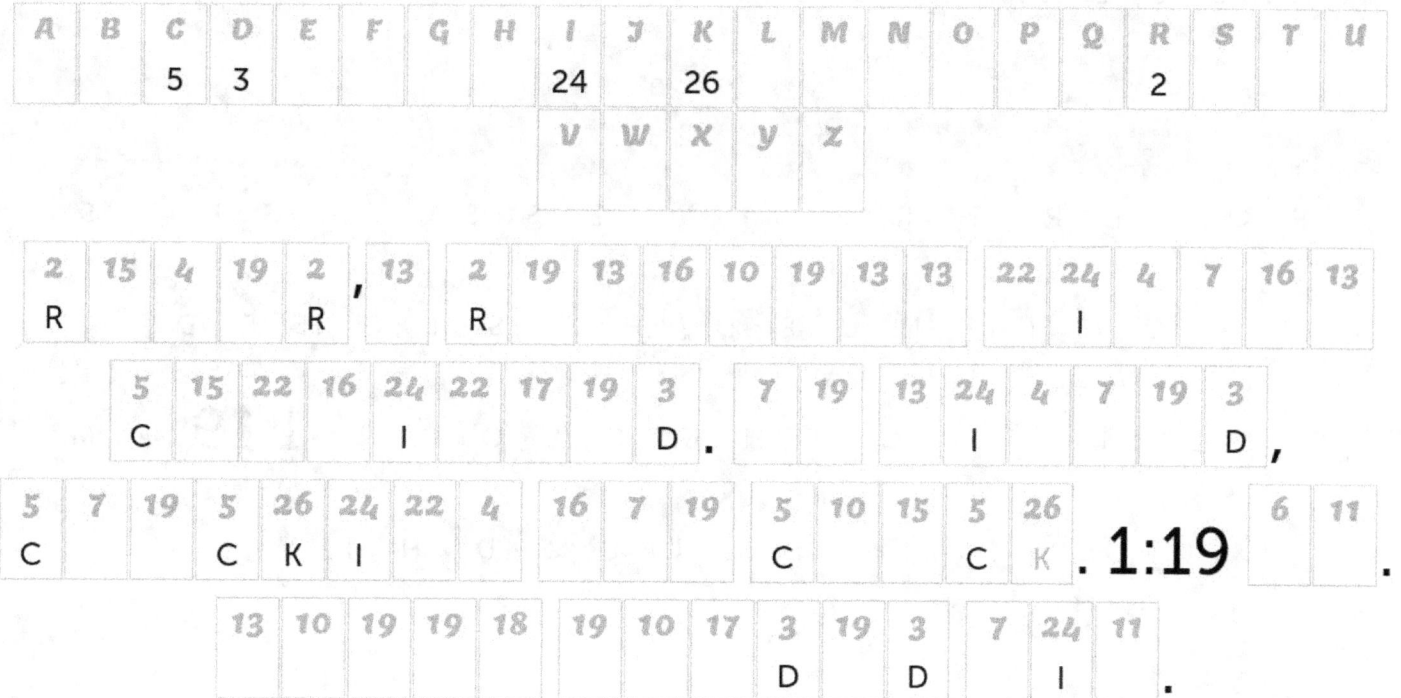

Answer.

ROGER'S RESTLESS NIGHTS CONTINUED. HE SIGHED, CHECKING THE CLOCK. 1:19 AM. SLEEP ELUDED HIM.

Quiz 4.   What is this secret message about?
Difficulty: * * *

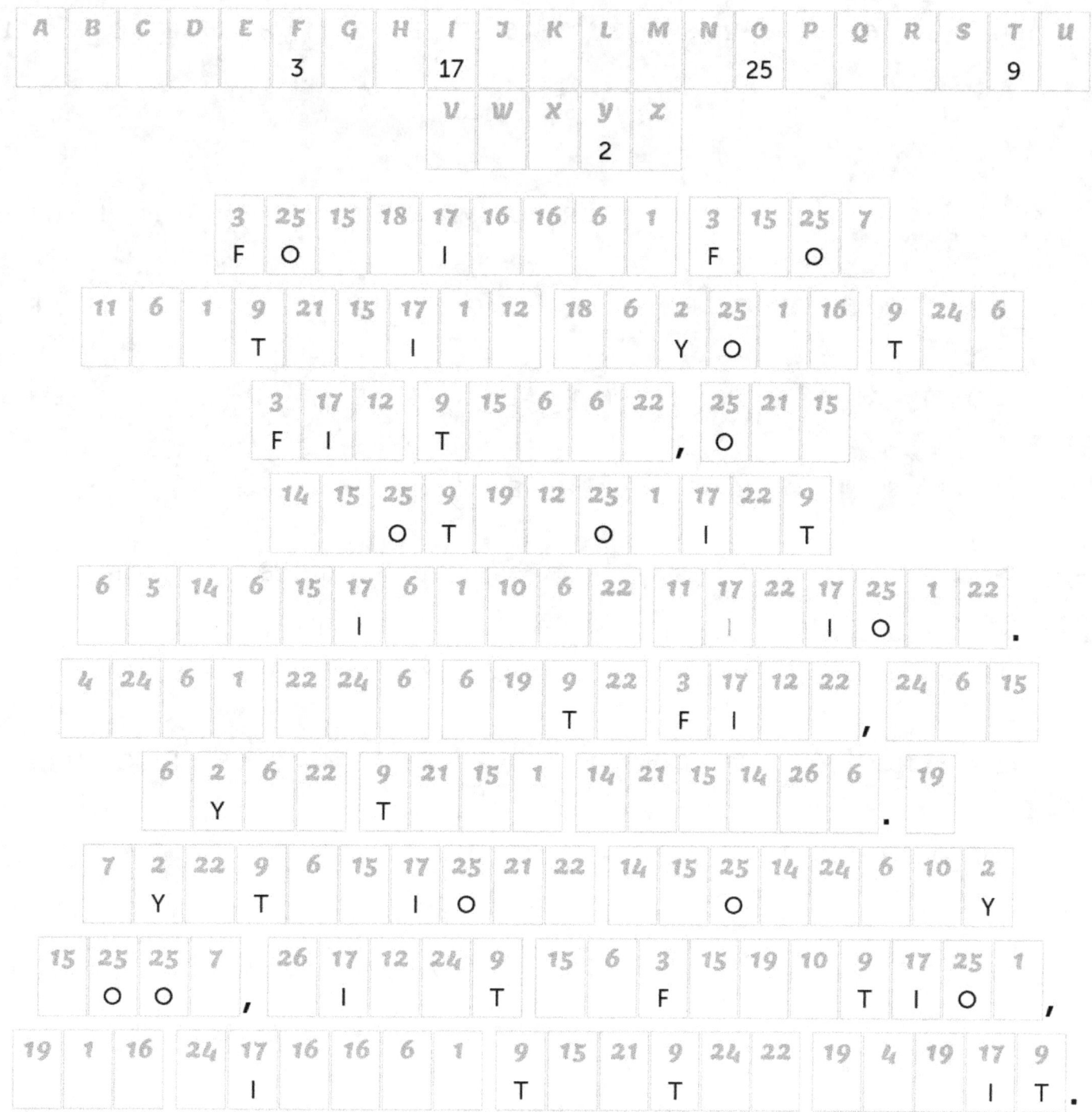

Answer.

A	B	C	D	E	F	G	H	I	J	K	L	M	N	O	P	Q	R	S	T	U	V
19	18	10	16	6	3	12	24	17	13	23	26	7	1	25	14	8	15	22	9	21	11

W	X	Y	Z
4	5	2	20

FORBIDDEN FROM VENTURING BEYOND THE FIG TREES, OUR PROTAGONIST EXPERIENCES VISIONS. WHEN SHE EATS FIGS, HER EYES TURN PURPLE. A MYSTERIOUS PROPHECY ROOM, LIGHT REFRACTION, AND HIDDEN TRUTHS AWAIT.

Quiz 5.   What is this secret message about?
Difficulty: * * *

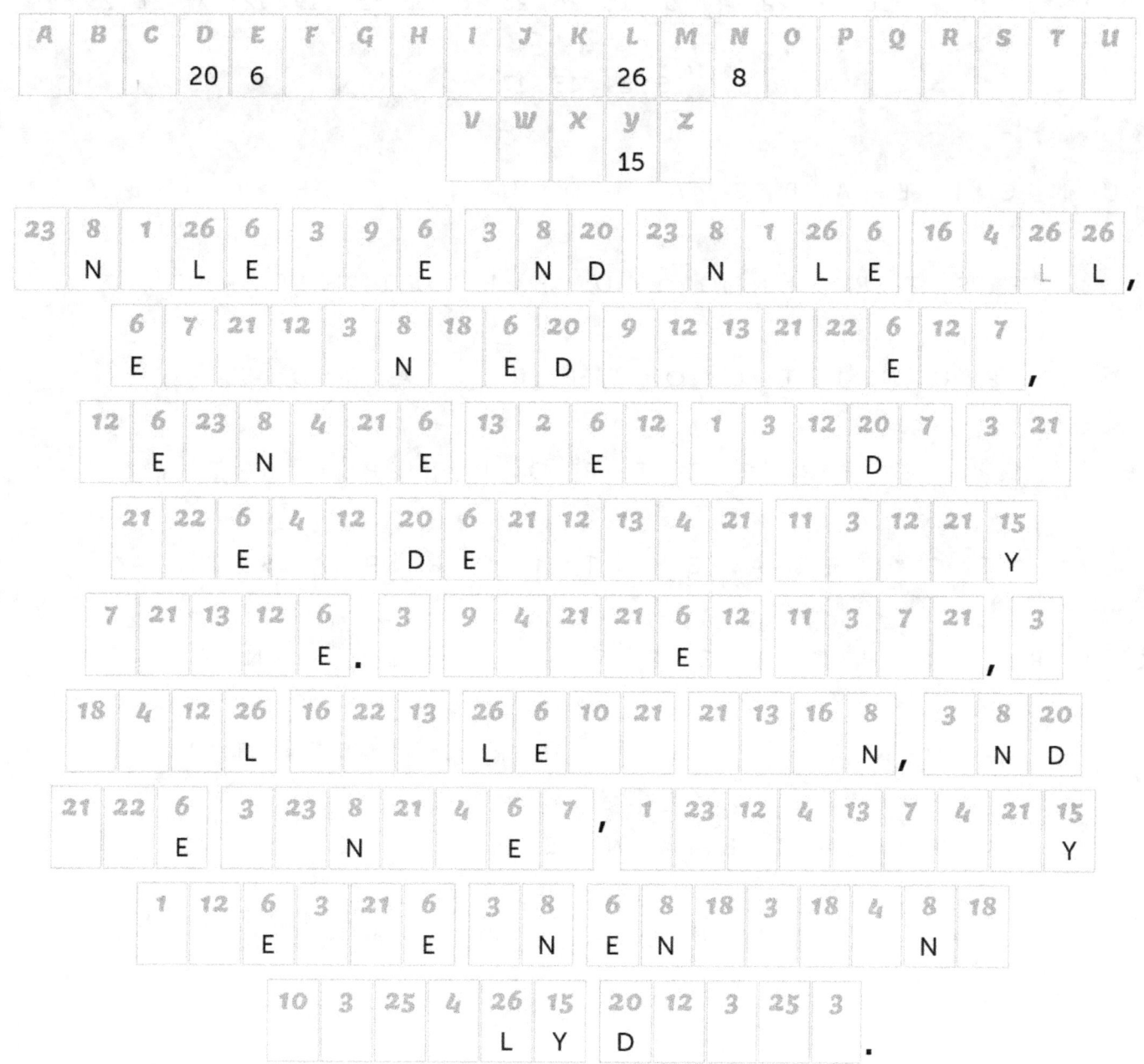

Answer.

A	B	C	D	E	F	G	H	I	J	K	L	M	N	O	P	Q	R	S	T	U	V
3	9	1	20	6	10	18	22	4	5	14	26	25	8	13	11	19	12	7	21	23	2

W	X	Y	Z
16	24	15	17

UNCLE ABE AND UNCLE WILL, ESTRANGED BROTHERS, REUNITE OVER CARDS AT THEIR DETROIT PARTY STORE. A BITTER PAST, A GIRL WHO LEFT TOWN, AND THE AUNTIES' CURIOSITY CREATE AN ENGAGING FAMILY DRAMA.

Quiz 6. What is this secret message about?
Difficulty: * * *

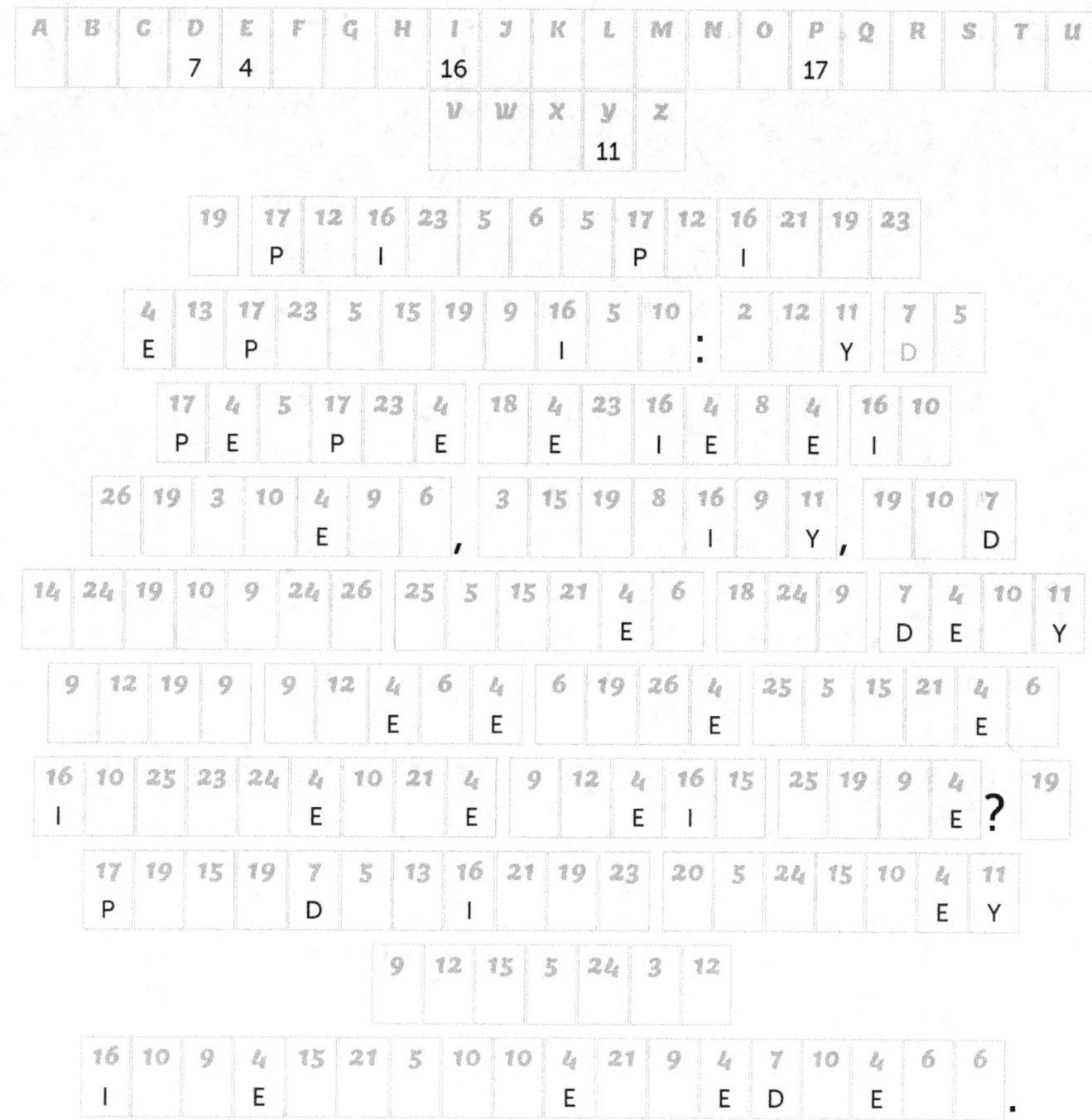

Answer.

A	B	C	D	E	F	G	H	I	J	K	L	M	N	O	P	Q	R	S	T	U	V
19	18	21	7	4	25	3	12	16	20	1	23	26	10	5	17	14	15	6	9	24	8

W	X	Y	Z
2	13	11	22

A PHILOSOPHICAL EXPLORATION: WHY DO PEOPLE BELIEVE IN MAGNETS, GRAVITY, AND QUANTUM FORCES BUT DENY THAT THESE SAME FORCES INFLUENCE THEIR FATE? A PARADOXICAL JOURNEY THROUGH INTERCONNECTEDNESS.

Quiz 7. What is this secret message about?
Difficulty: * * *

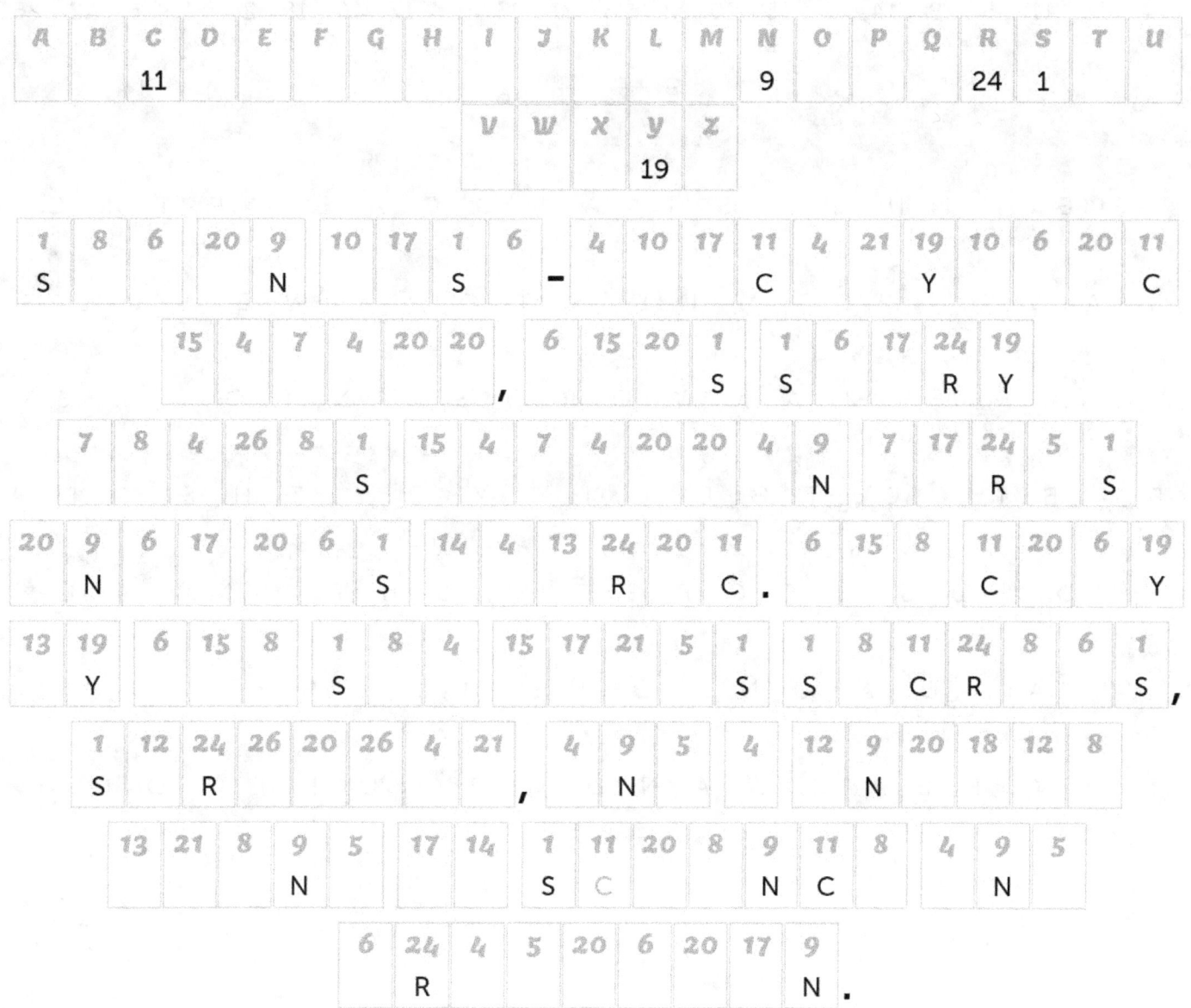

Answer.

SET IN POST-APOCALYPTIC HAWAII, THIS STORY WEAVES HAWAIIAN WORDS INTO ITS FABRIC. THE CITY BY THE SEA HOLDS SECRETS, SURVIVAL, AND A UNIQUE BLEND OF SCIENCE AND TRADITION.

Quiz 8. What is this secret message about?
Difficulty: * * *

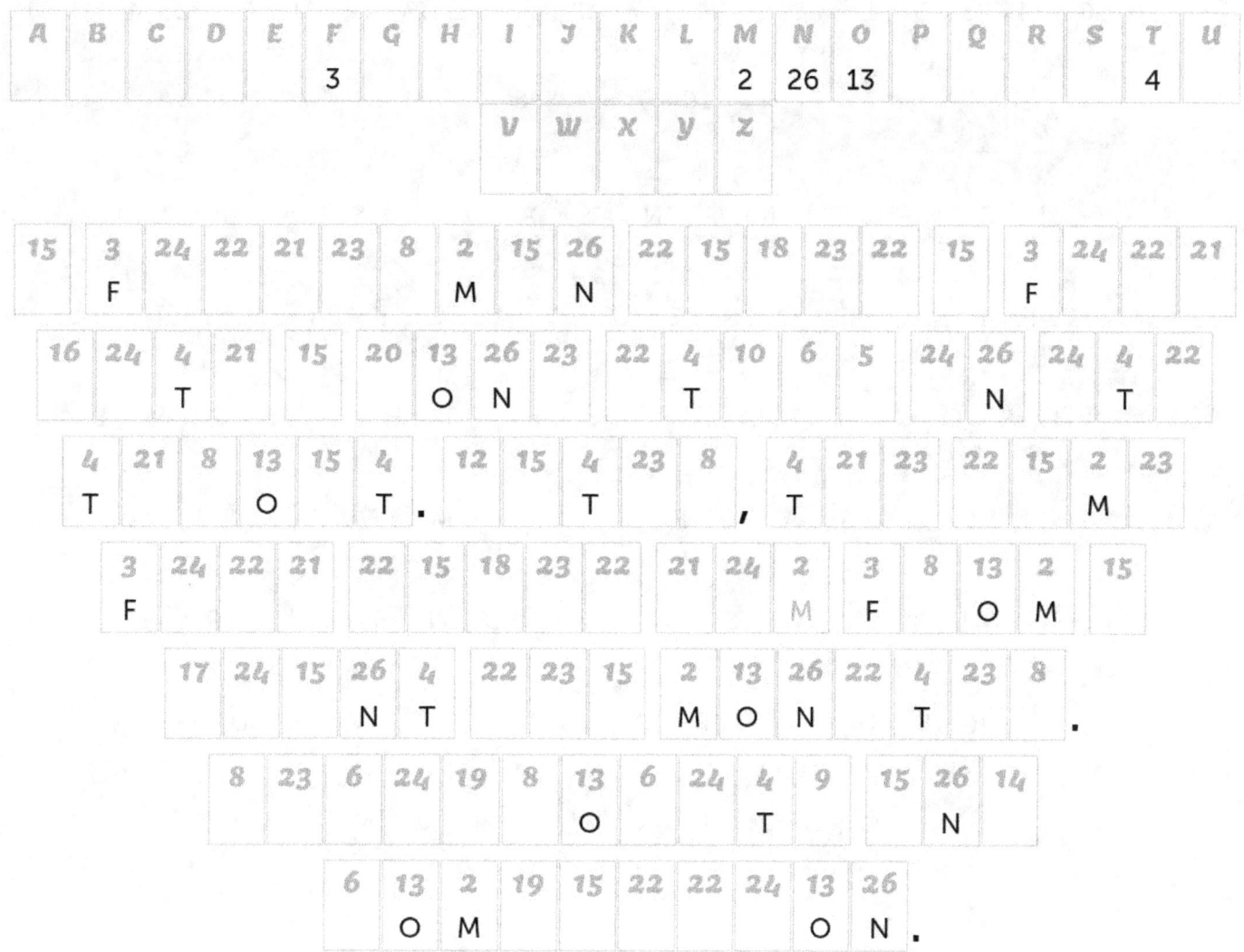

Answer.

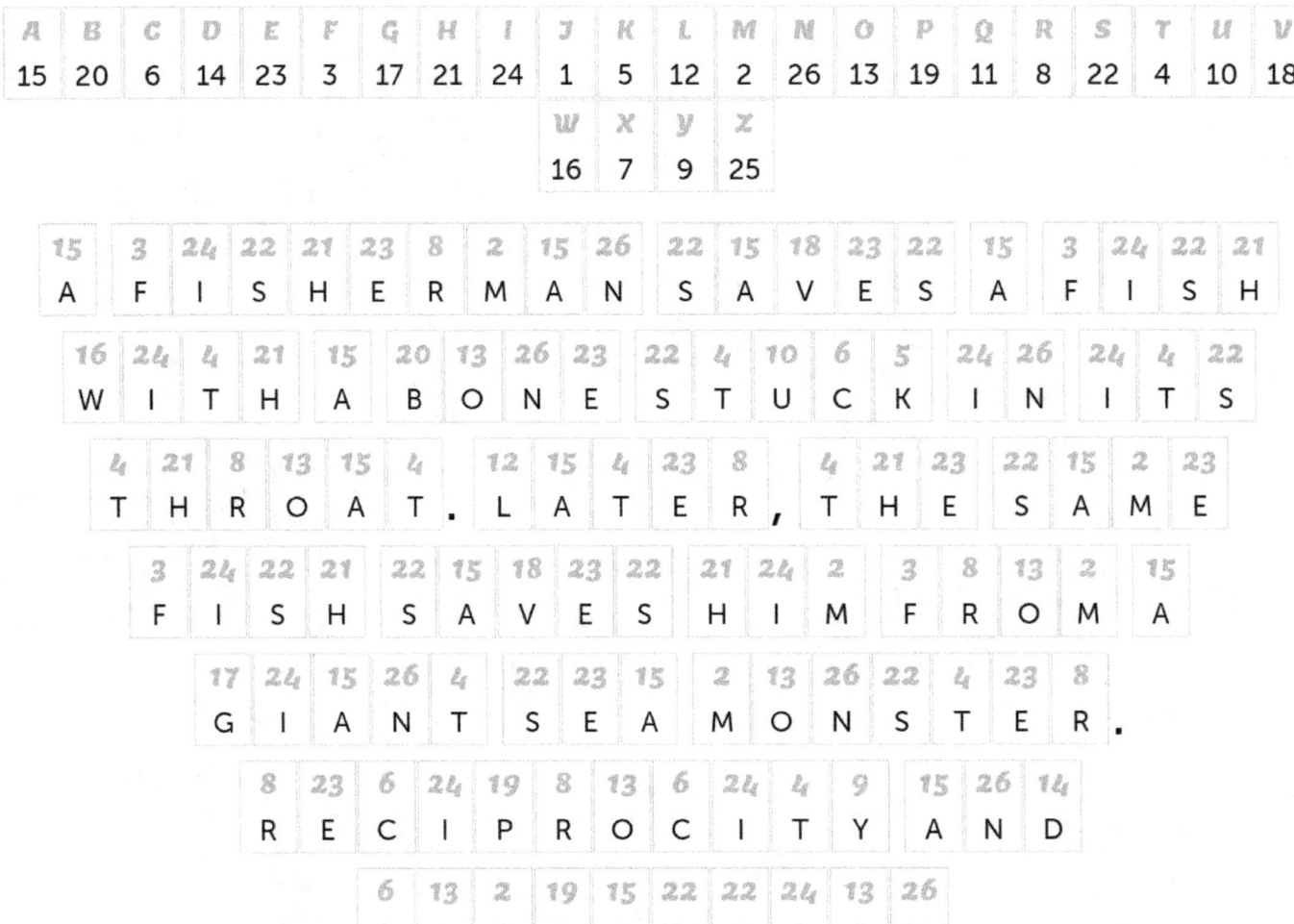

Quiz 9. What is this secret message about?
Difficulty: * * * *

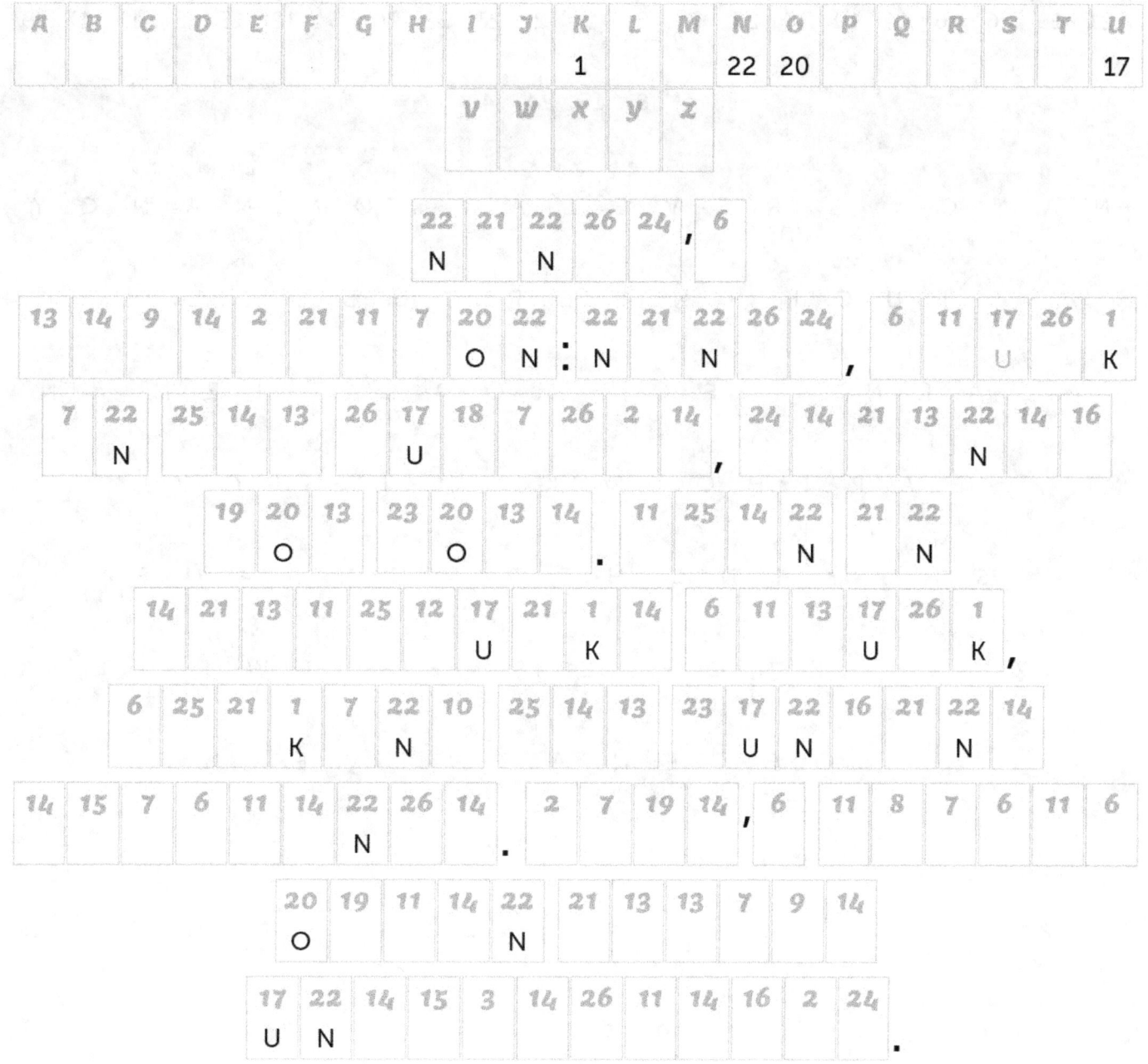

Answer.

A	B	C	D	E	F	G	H	I	J	K	L	M	N	O	P	Q	R	S	T	U	V
21	18	26	16	14	19	10	25	7	5	1	2	23	22	20	3	12	13	6	11	17	9

W	X	Y	Z
8	15	24	4

NANCY'S REVELATION: NANCY, STUCK IN HER CUBICLE, YEARNED FOR MORE. THEN AN EARTHQUAKE STRUCK, SHAKING HER MUNDANE EXISTENCE. LIFE'S TWISTS OFTEN ARRIVE UNEXPECTEDLY.

Quiz 10. What is this secret message about?
Difficulty: * * * *

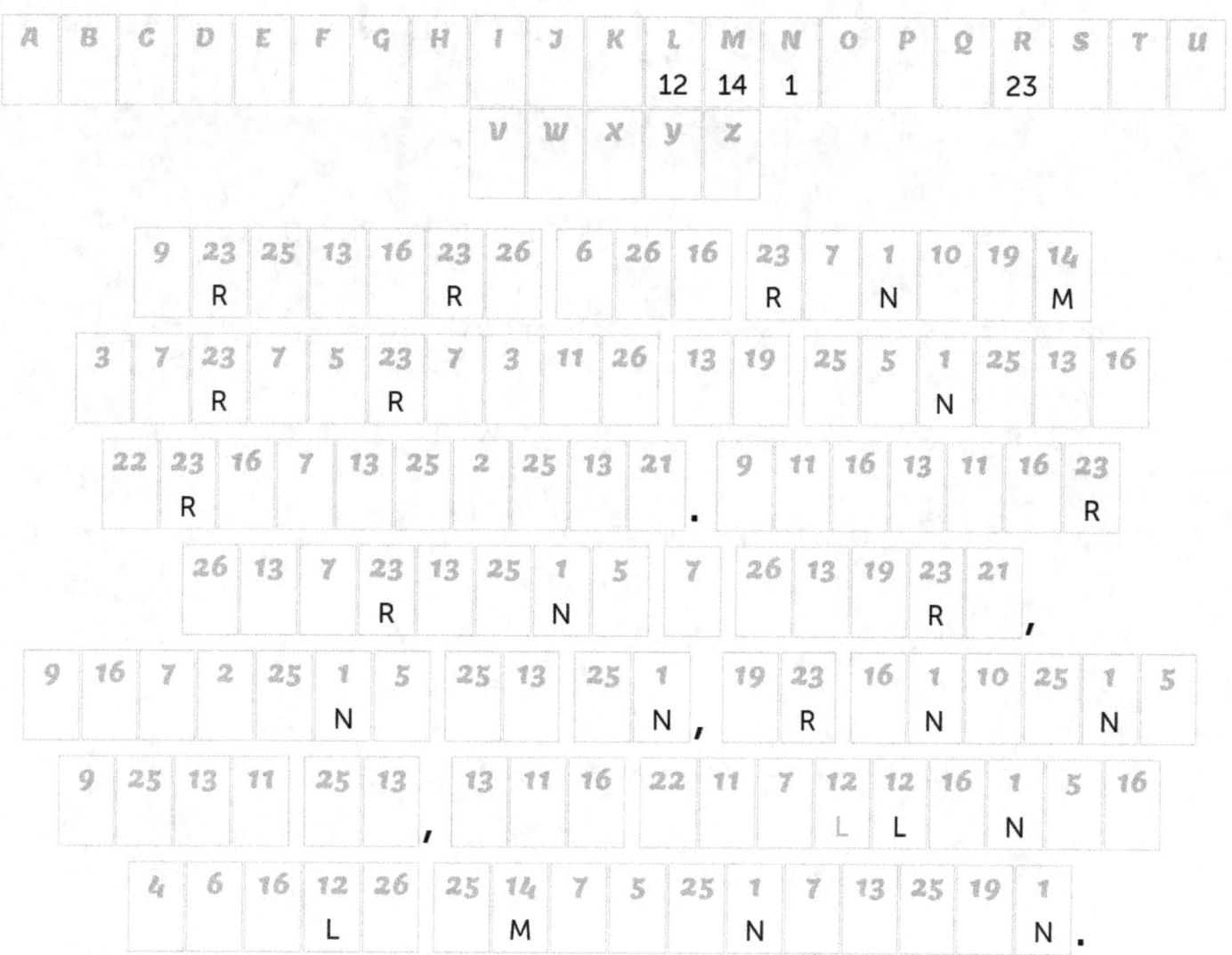

Answer.

WRITERS USE RANDOM PARAGRAPHS TO IGNITE CREATIVITY. WHETHER STARTING A STORY, WEAVING IT IN, OR ENDING WITH IT, THE CHALLENGE FUELS IMAGINATION.

Quiz 11.   What is this secret message about?
Difficulty: * * * *

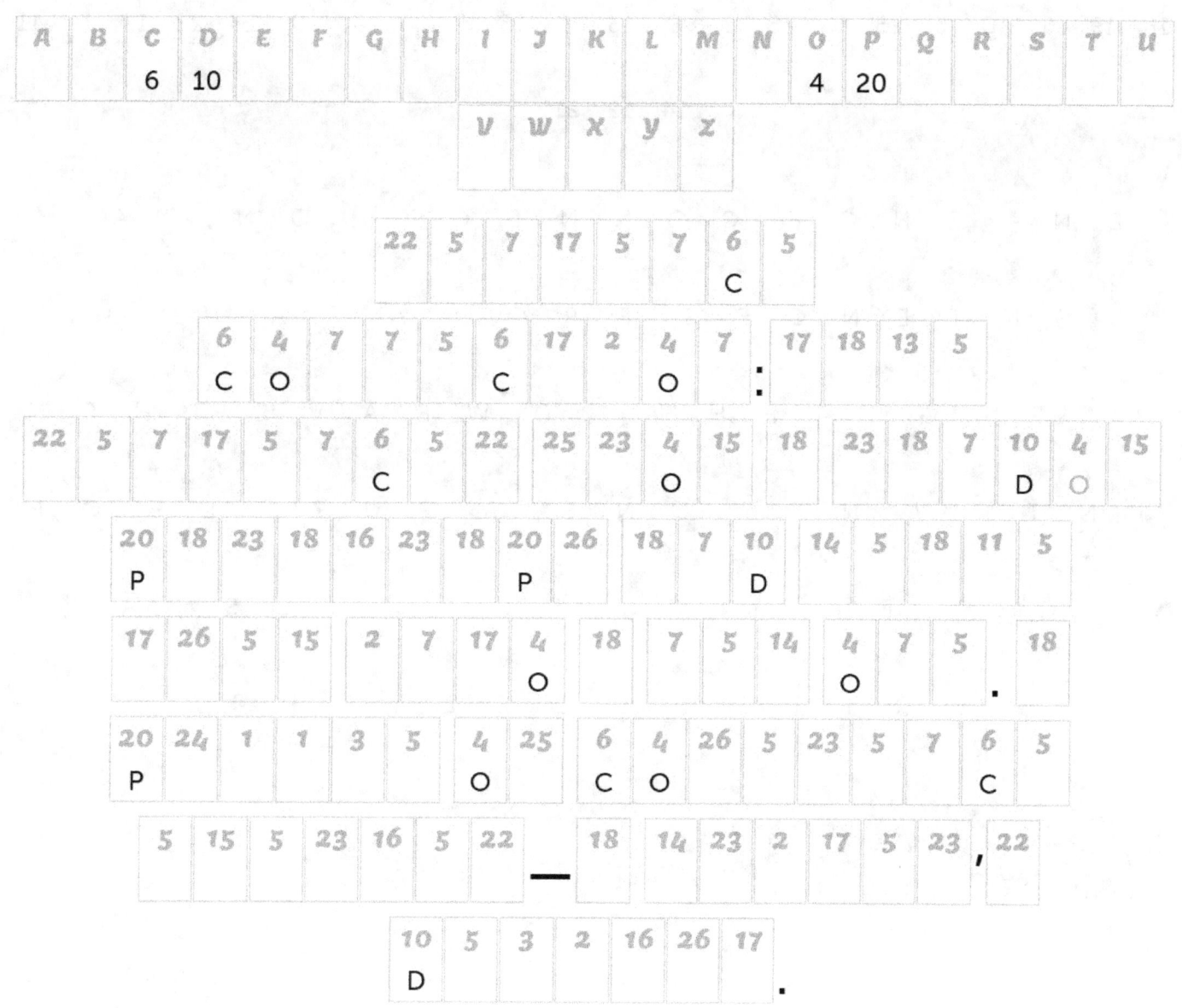

Answer.

A	B	C	D	E	F	G	H	I	J	K	L	M	N	O	P	Q	R	S	T	U	V
18	19	6	10	5	25	16	26	2	8	13	3	15	7	4	20	9	23	22	17	24	11

W	X	Y	Z
14	12	21	1

SENTENCE CONNECTION: TAKE SENTENCES FROM A RANDOM PARAGRAPH AND WEAVE THEM INTO A NEW ONE. A PUZZLE OF COHERENCE EMERGES—A WRITER'S DELIGHT.

Quiz 12. What is this secret message about?
Difficulty: * * * *

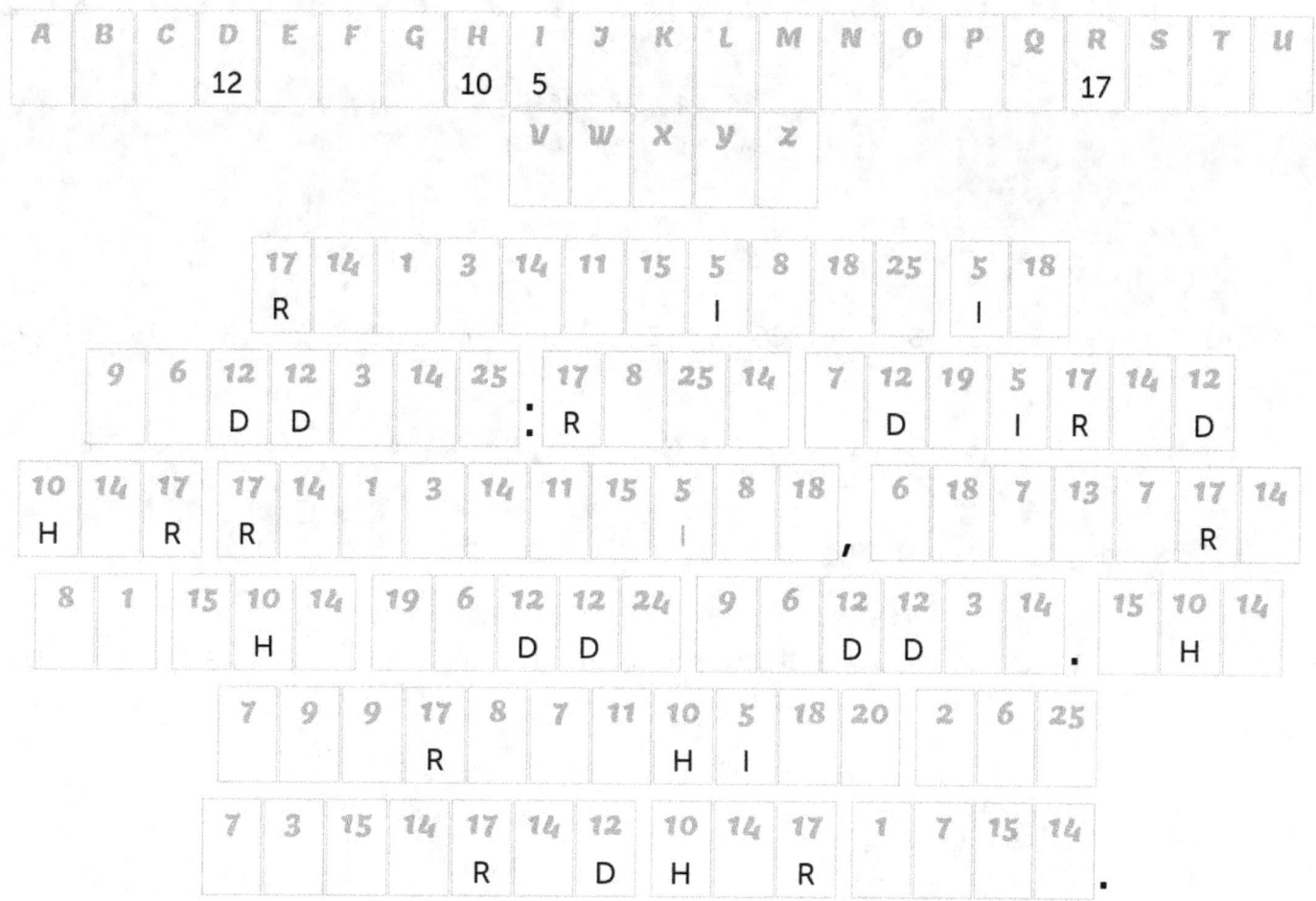

Answer.

REFLECTIONS IN PUDDLES: ROSE ADMIRED HER REFLECTION, UNAWARE OF THE MUDDY PUDDLE. THE APPROACHING BUS ALTERED HER FATE.

Quiz 13. What is this secret message about?
Difficulty: * * * *

Answer.

FIFI'S DISAPPEARANCE: JENNIFER'S CAT VANISHED. GUS THE DOG SEEMED PLEASED. WAS HE HIDING SOMETHING?

Quiz 14. What is this secret message about?
Difficulty: * * * *

Answer.

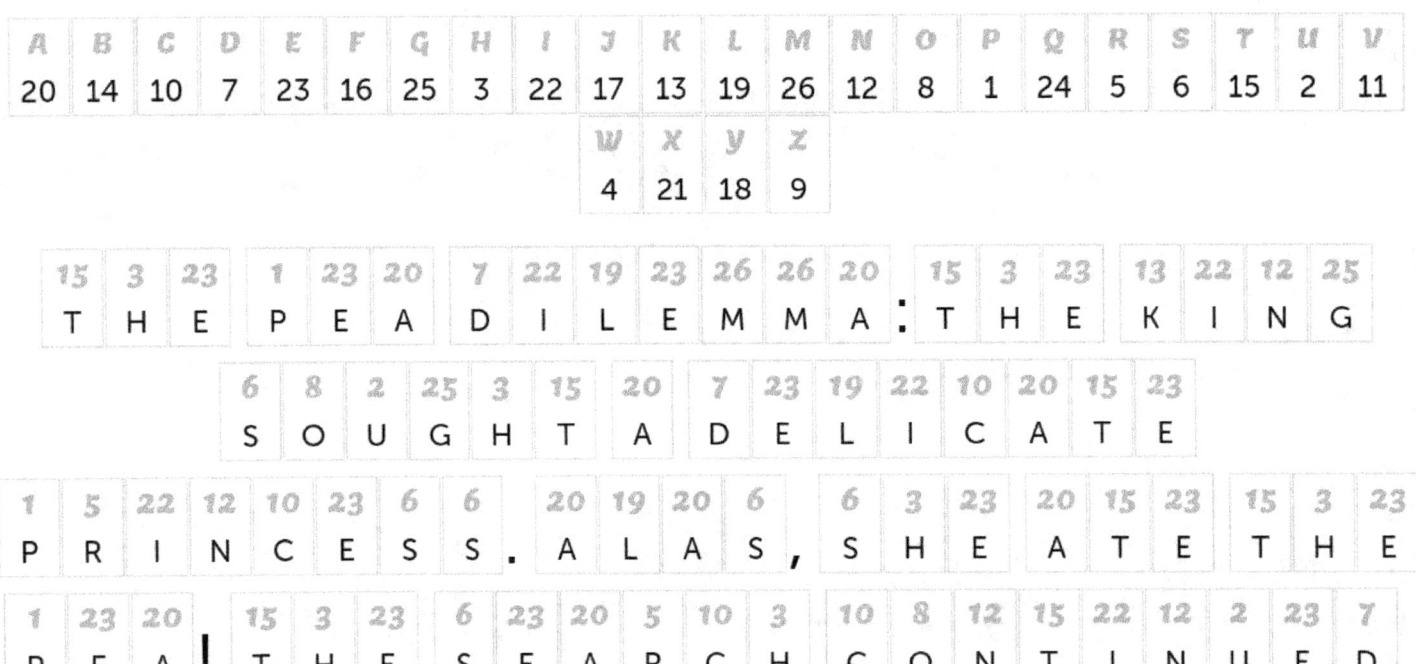

THE PEA DILEMMA: THE KING SOUGHT A DELICATE PRINCESS. ALAS, SHE ATE THE PEA! THE SEARCH CONTINUED.

Quiz 15. What is this secret message about?
Difficulty: * * * *

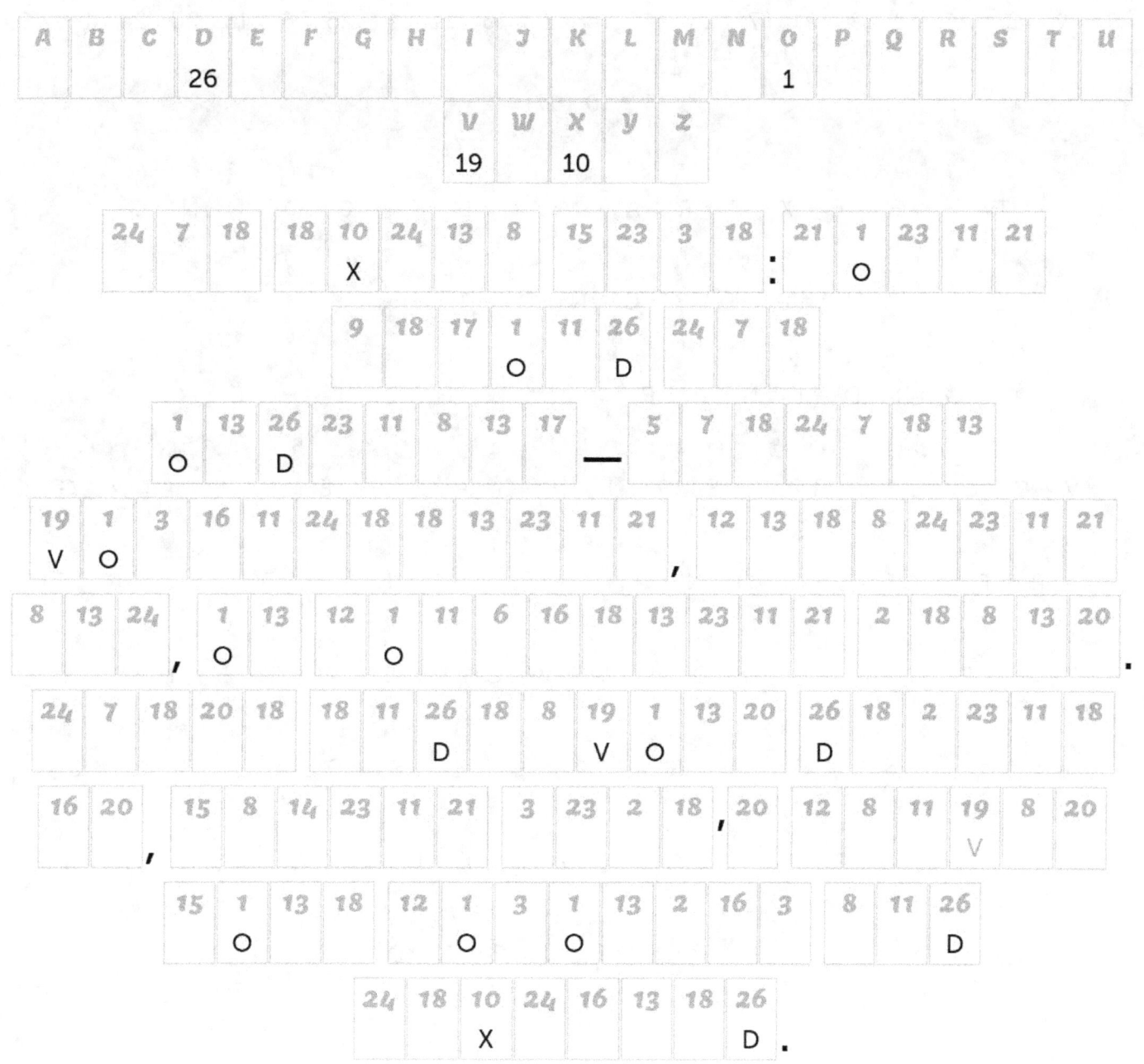

Answer.

A	B	C	D	E	F	G	H	I	J	K	L	M	N	O	P	Q	R	S	T	U	V
8	9	12	26	18	2	21	7	23	4	14	3	15	11	1	25	6	13	20	24	16	19

W	X	Y	Z
5	10	17	22

THE EXTRA MILE: GOING BEYOND THE ORDINARY — WHETHER VOLUNTEERING, CREATING ART, OR CONQUERING FEARS. THESE ENDEAVORS DEFINE US, MAKING LIFE'S CANVAS MORE COLORFUL AND TEXTURED.

Quiz 16. What is this secret message about?
Difficulty: * * * *

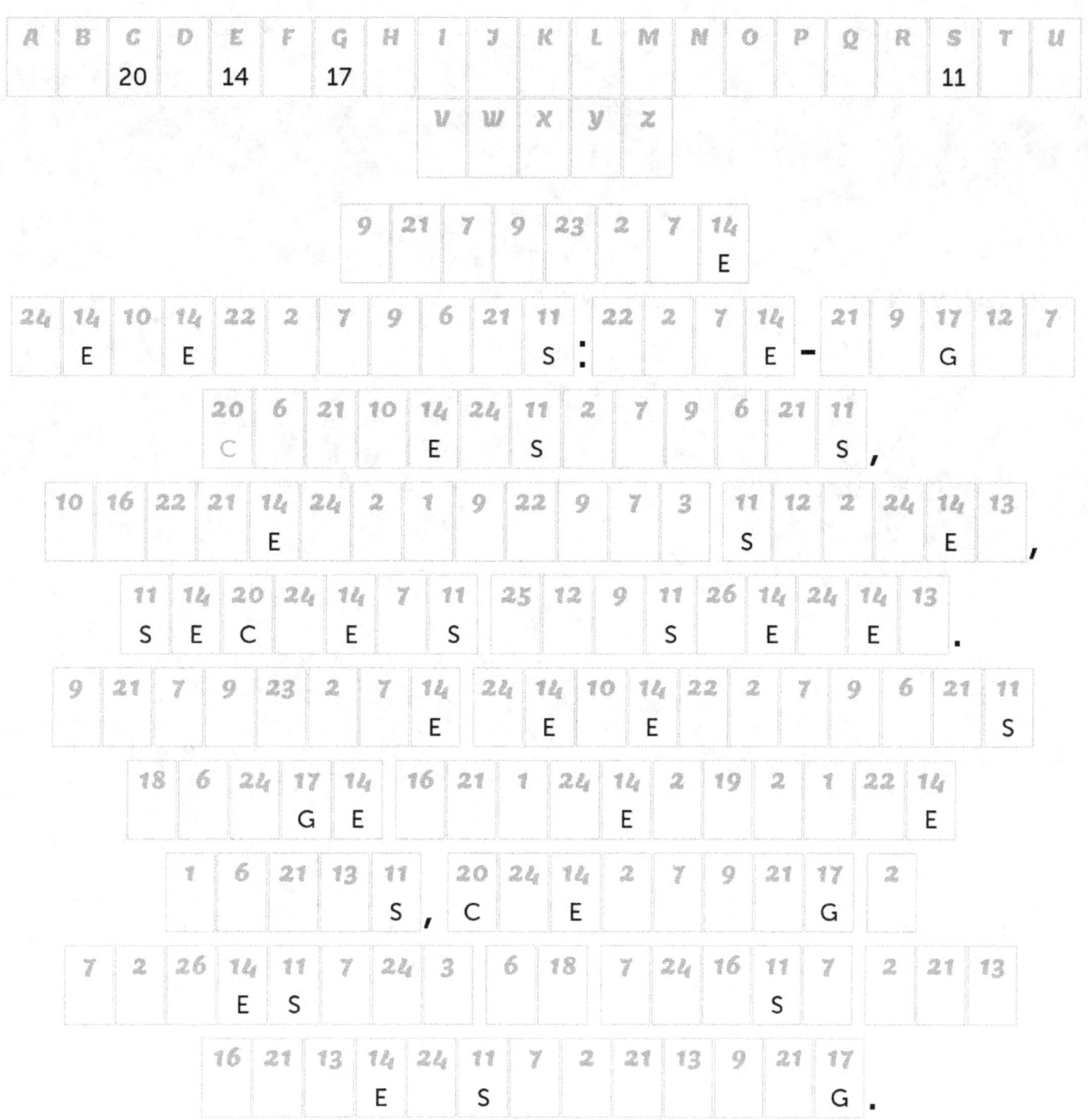

Answer.

A	B	C	D	E	F	G	H	I	J	K	L	M	N	O	P	Q	R	S	T	U	V
2	1	20	13	14	18	17	12	9	15	19	22	23	21	6	26	4	24	11	7	16	10

W	X	Y	Z
25	5	3	8

INTIMATE REVELATIONS: LATE-NIGHT CONVERSATIONS, VULNERABILITY SHARED, SECRETS WHISPERED. INTIMATE REVELATIONS FORGE UNBREAKABLE BONDS, CREATING A TAPESTRY OF TRUST AND UNDERSTANDING.

Quiz 17. What is this secret message about?
Difficulty: * * * *

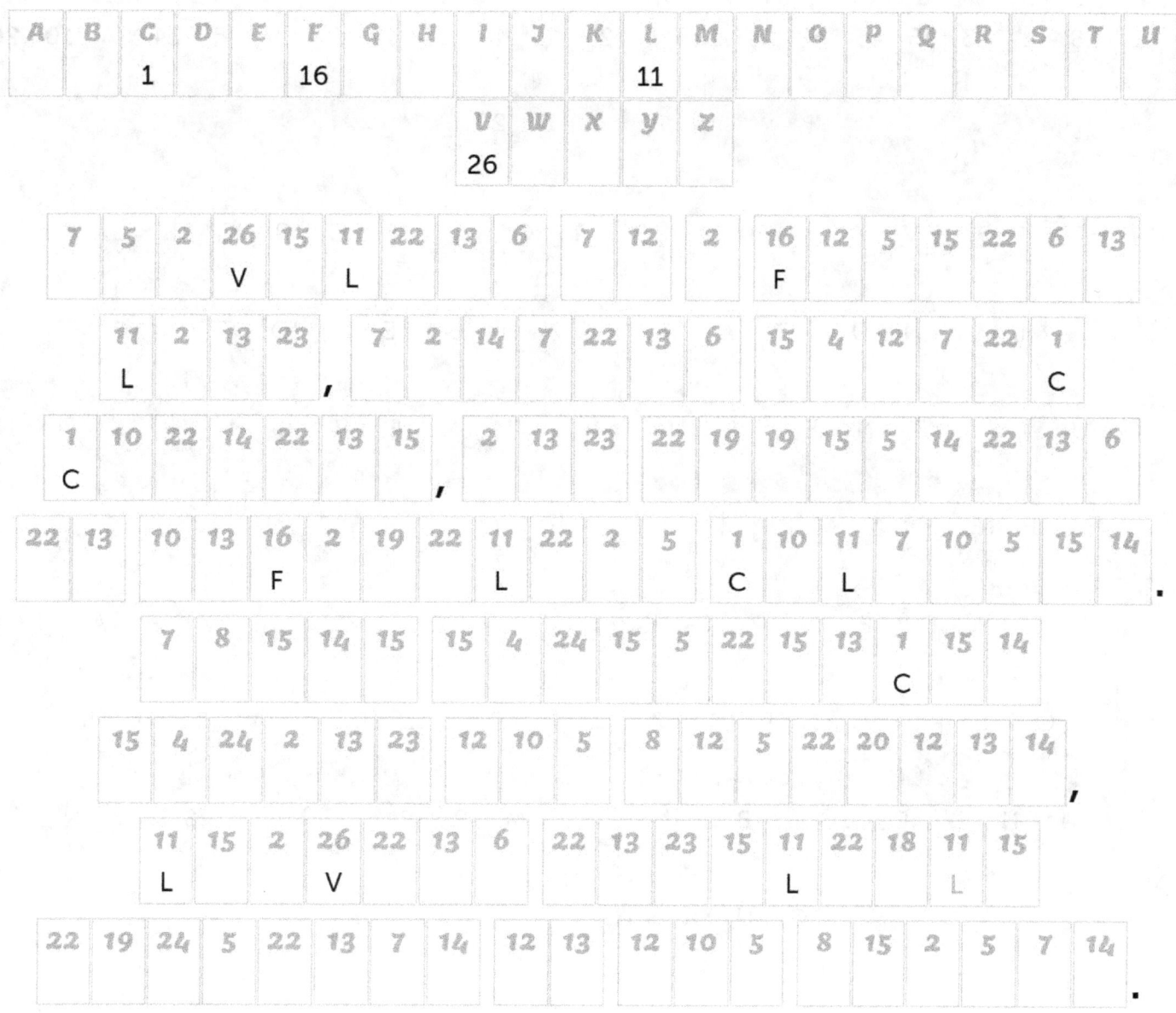

Answer.

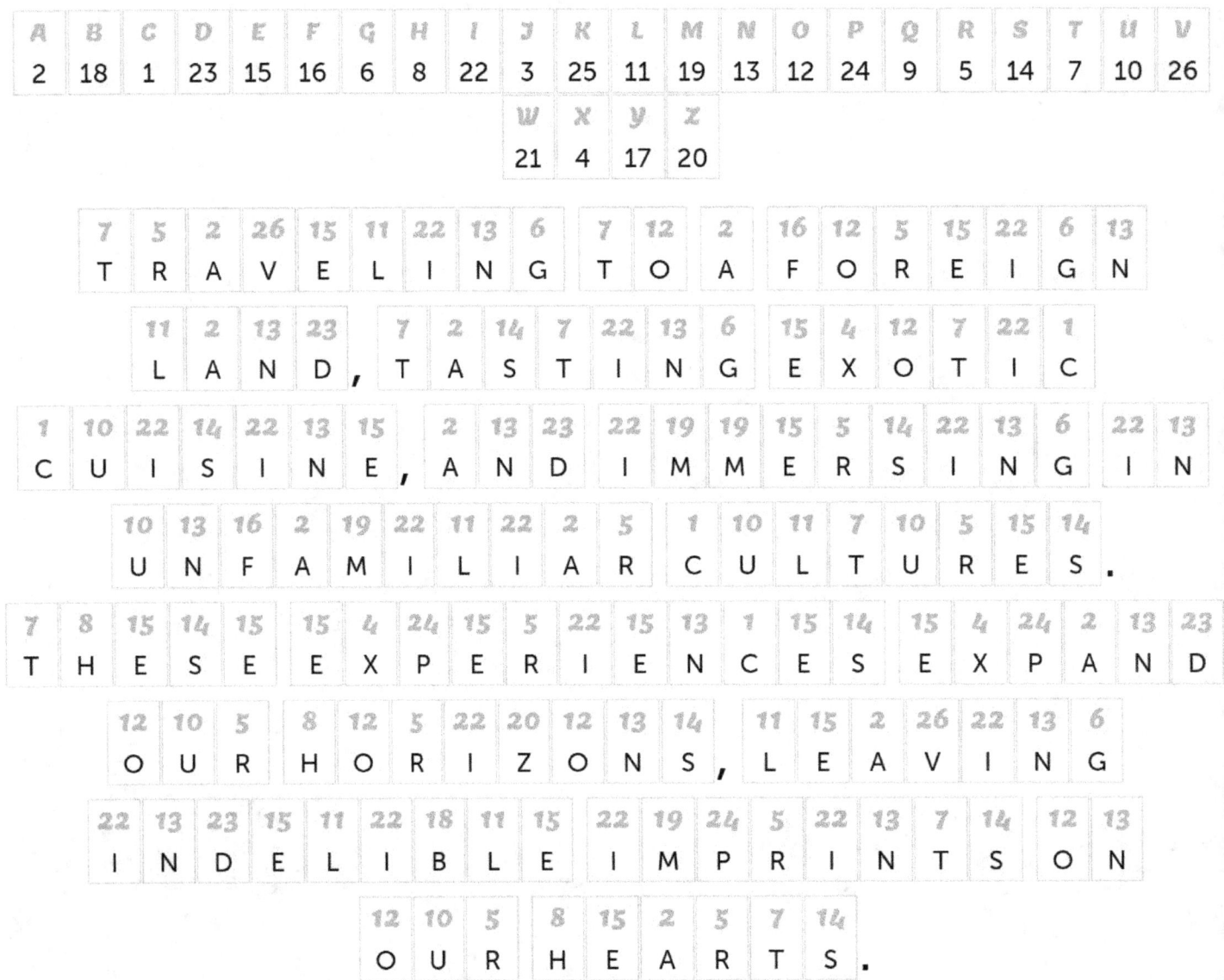

TRAVELING TO A FOREIGN LAND, TASTING EXOTIC CUISINE, AND IMMERSING IN UNFAMILIAR CULTURES. THESE EXPERIENCES EXPAND OUR HORIZONS, LEAVING INDELIBLE IMPRINTS ON OUR HEARTS.

Quiz 18. What is this secret message about?
Difficulty: * * * *

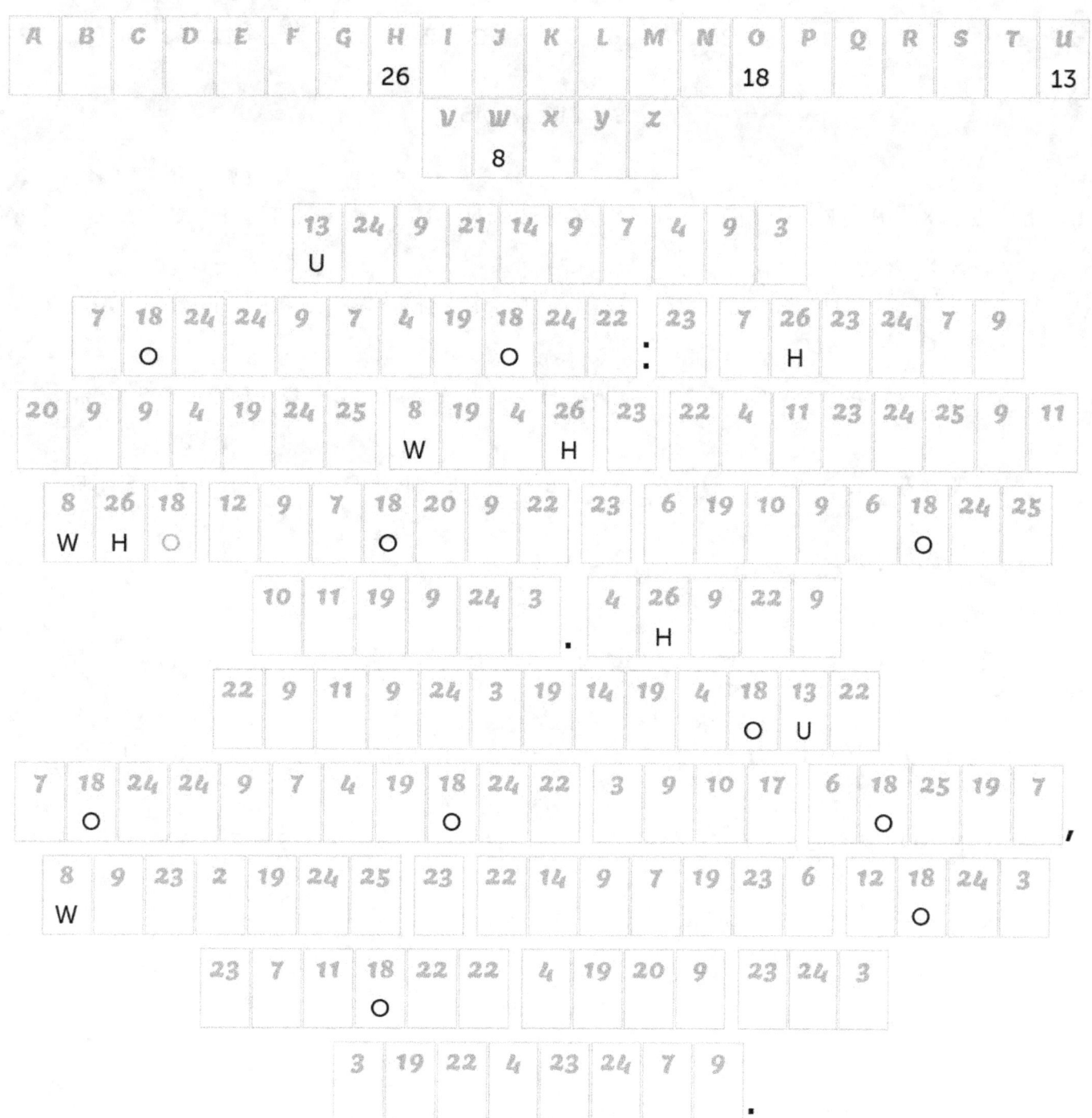

Answer.

A	B	C	D	E	F	G	H	I	J	K	L	M	N	O	P	Q	R	S	T	U	V
23	12	7	3	9	10	25	26	19	16	5	6	20	24	18	14	1	11	22	4	13	2

W	X	Y	Z
8	21	17	15

UNEXPECTED CONNECTIONS: A CHANCE MEETING WITH A STRANGER WHO BECOMES A LIFELONG FRIEND. THESE SERENDIPITOUS CONNECTIONS DEFY LOGIC, WEAVING A SPECIAL BOND ACROSS TIME AND DISTANCE.

Quiz 19. What is this secret message about?
Difficulty: * * * *

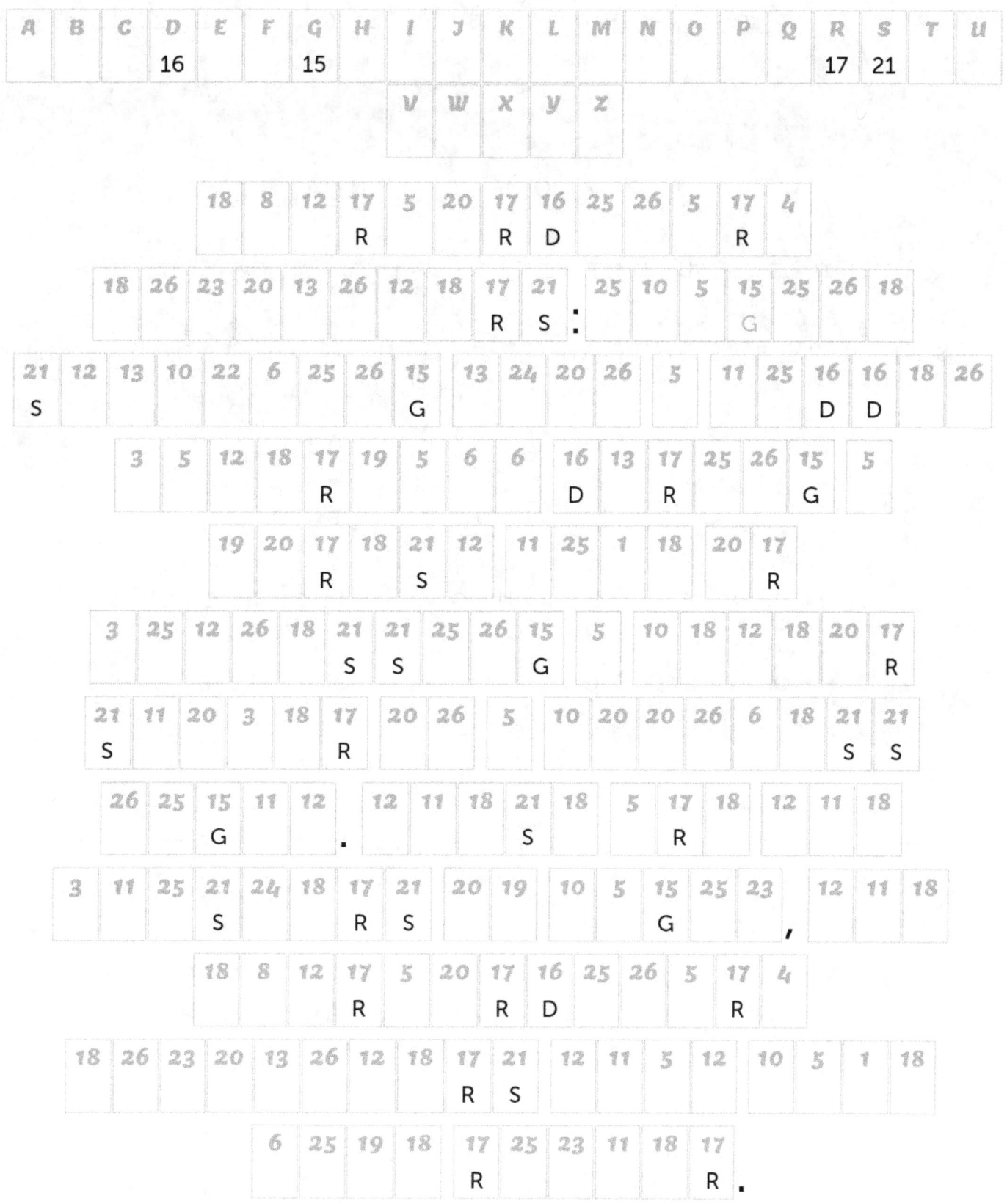

Answer.

A	B	C	D	E	F	G	H	I	J	K	L	M	N	O	P	Q	R	S	T	U	V
5	22	23	16	18	19	15	11	25	7	1	6	10	26	20	24	9	17	21	12	13	2

W	X	Y	Z
3	8	4	14

EXTRAORDINARY ENCOUNTERS: IMAGINE STUMBLING UPON A HIDDEN WATERFALL DURING A FOREST HIKE OR WITNESSING A METEOR SHOWER ON A MOONLESS NIGHT. THESE ARE THE WHISPERS OF MAGIC, THE EXTRAORDINARY ENCOUNTERS THAT MAKE LIFE RICHER.

Quiz 20. What is this secret message about?
Difficulty: * * * *

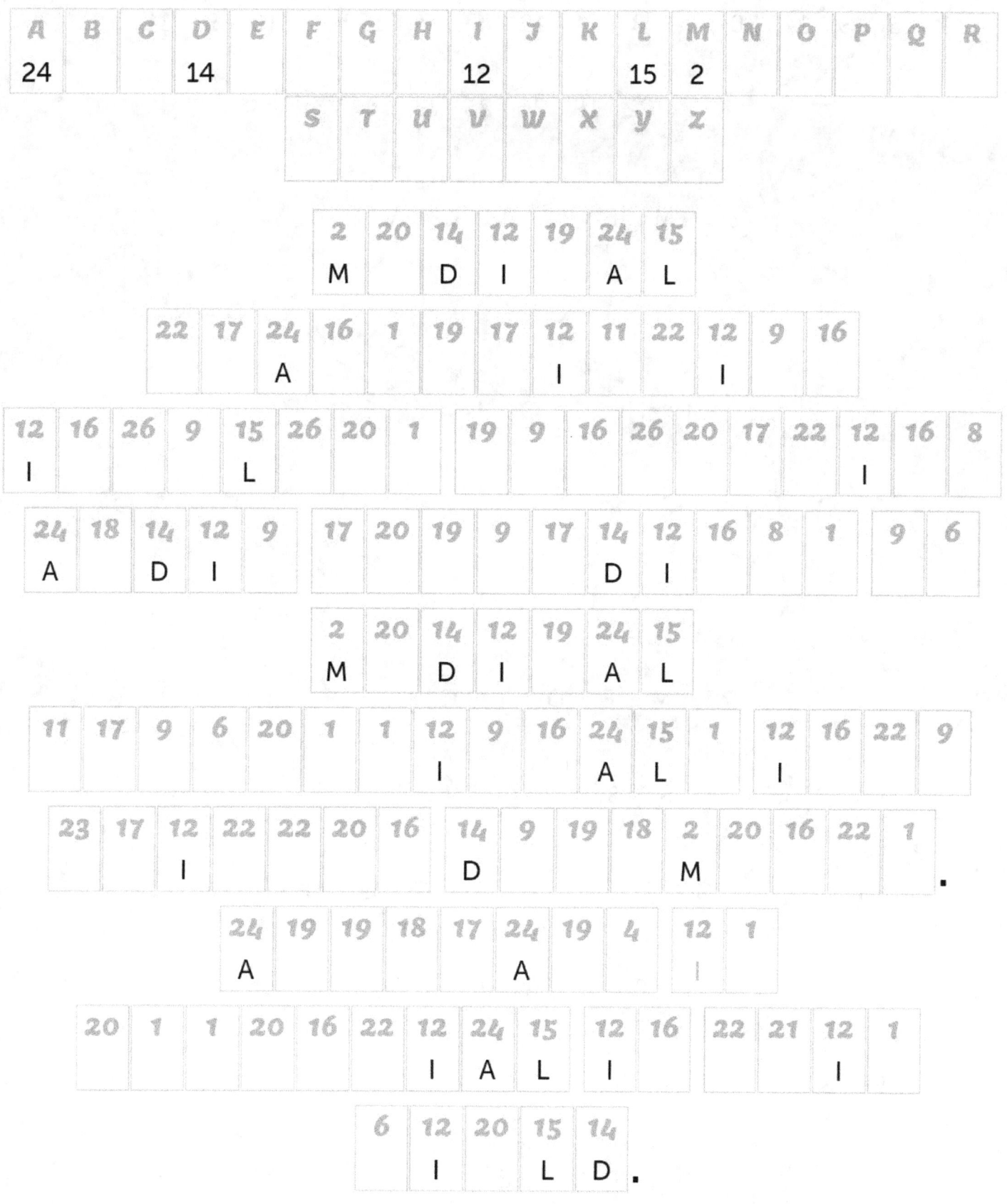

Answer.

A	B	C	D	E	F	G	H	I	J	K	L	M	N	O	P	Q	R
24	25	19	14	20	6	8	21	12	5	7	15	2	16	9	11	10	17

S	T	U	V	W	X	Y	Z
1	22	18	26	23	13	4	3

MEDICAL TRANSCRIPTION INVOLVES CONVERTING AUDIO RECORDINGS OF MEDICAL PROFESSIONALS INTO WRITTEN DOCUMENTS. ACCURACY IS ESSENTIAL IN THIS FIELD.

Quiz 21. What is this secret message about?
Difficulty: * * * *

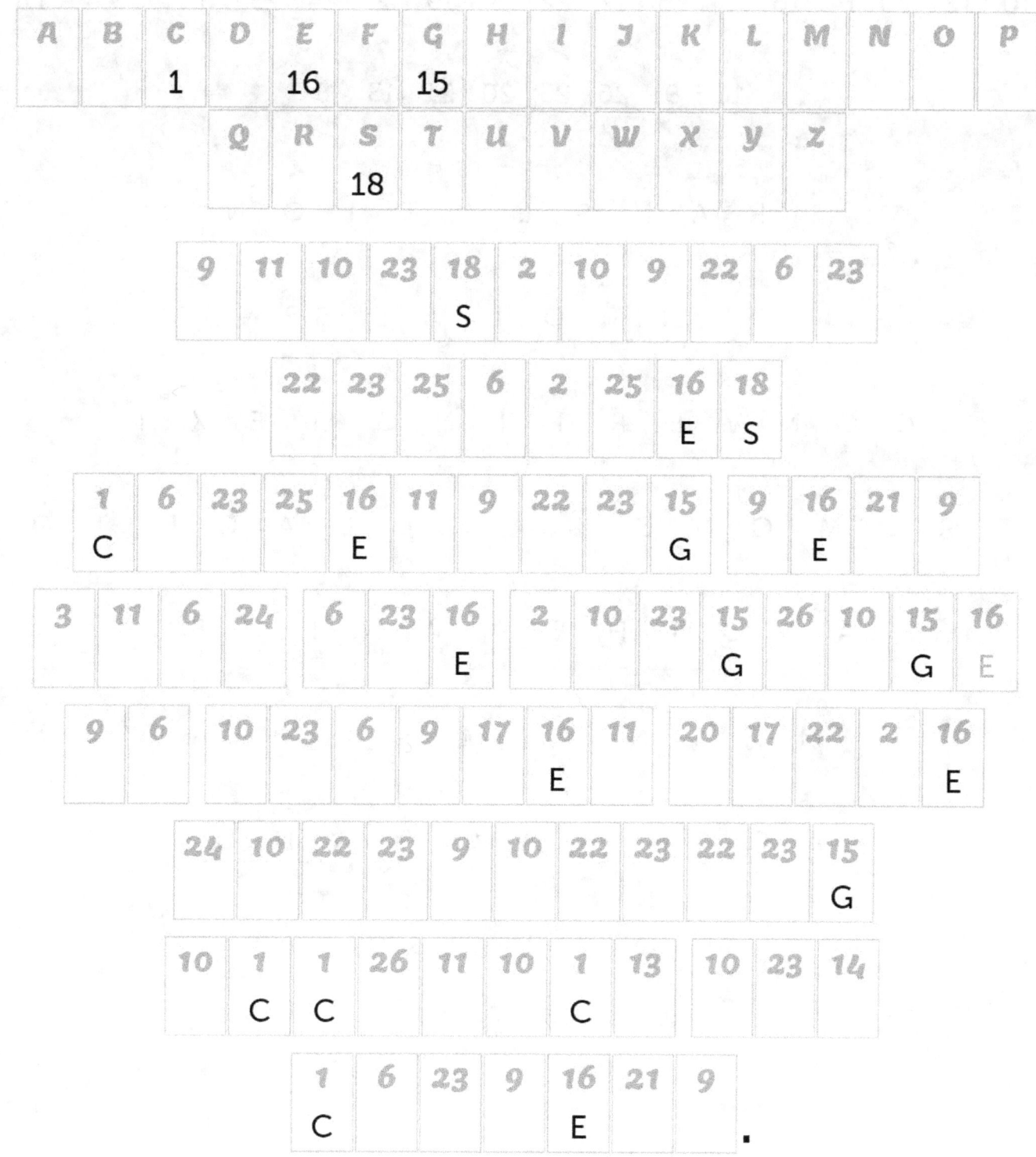

Answer.

TRANSLATION INVOLVES CONVERTING TEXT FROM ONE LANGUAGE TO ANOTHER WHILE MAINTAINING ACCURACY AND CONTEXT.

Quiz 22. What is this secret message about?
Difficulty: * * * *

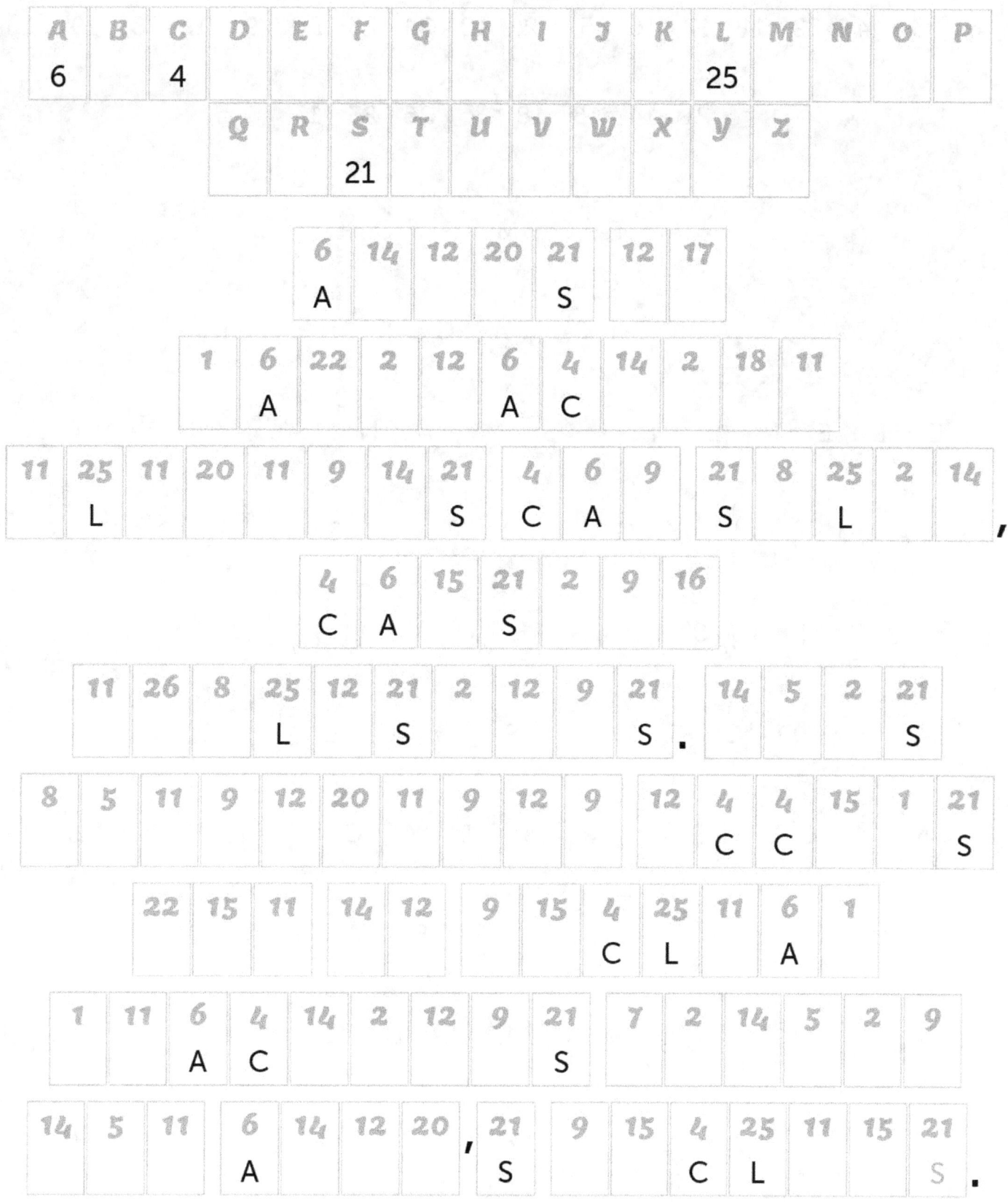

Answer.

ATOMS OF RADIOACTIVE ELEMENTS CAN SPLIT, CAUSING EXPLOSIONS. THIS PHENOMENON OCCURS DUE TO NUCLEAR REACTIONS WITHIN THE ATOM'S NUCLEUS.

Quiz 23. What is this secret message about?
Difficulty: * * * *

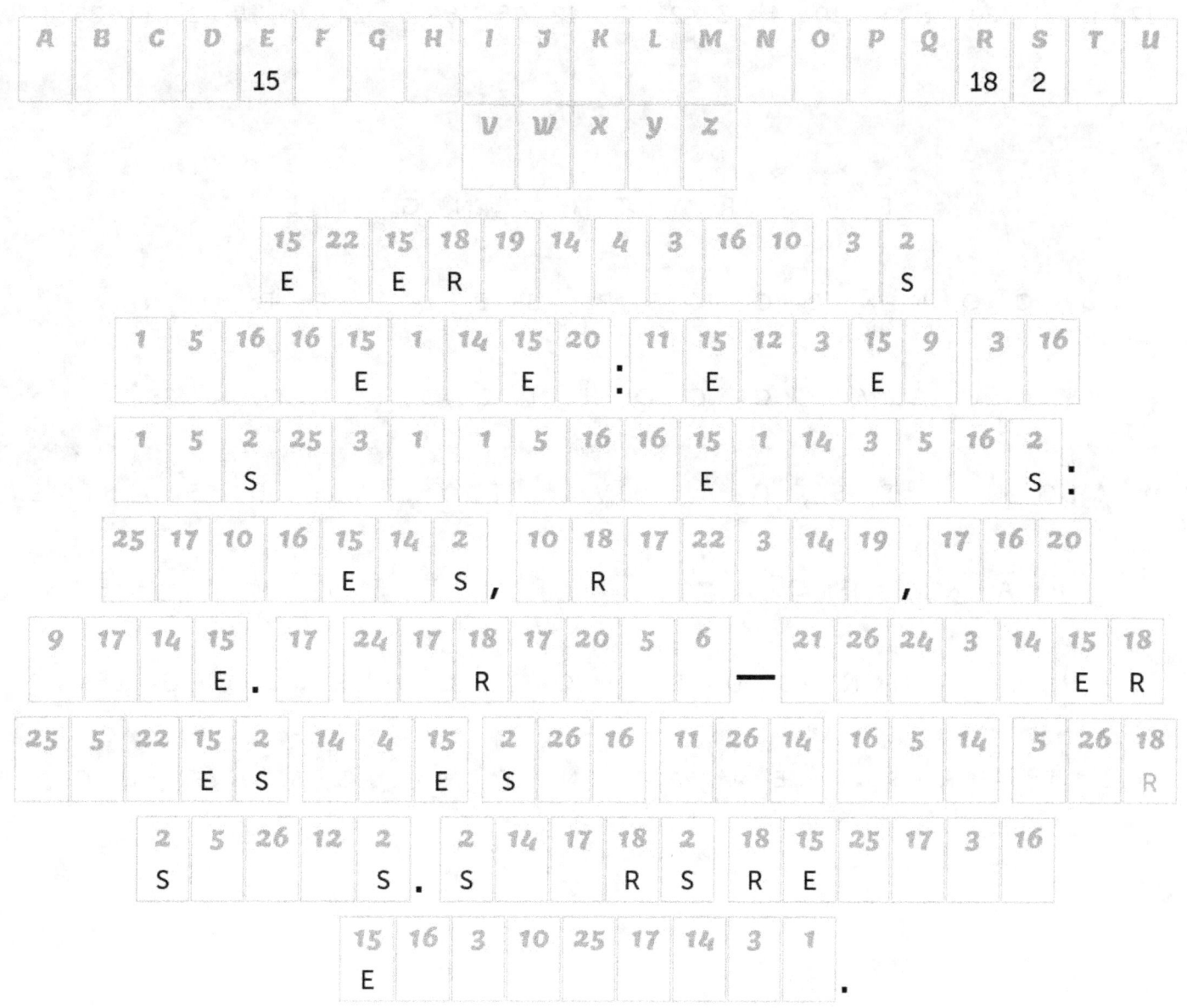

Answer.

A	B	C	D	E	F	G	H	I	J	K	L	M	N	O	P	Q	R	S	T	U	V
17	11	1	20	15	9	10	4	3	21	13	12	25	16	5	24	8	18	2	14	26	22

W	X	Y	Z
7	6	19	23

EVERYTHING IS CONNECTED: BELIEF IN COSMIC CONNECTIONS: MAGNETS, GRAVITY, AND FATE. A PARADOX — JUPITER MOVES THE SUN BUT NOT OUR SOULS. STARS REMAIN ENIGMATIC.

Quiz 24. What is this secret message about?
Difficulty: * * * *

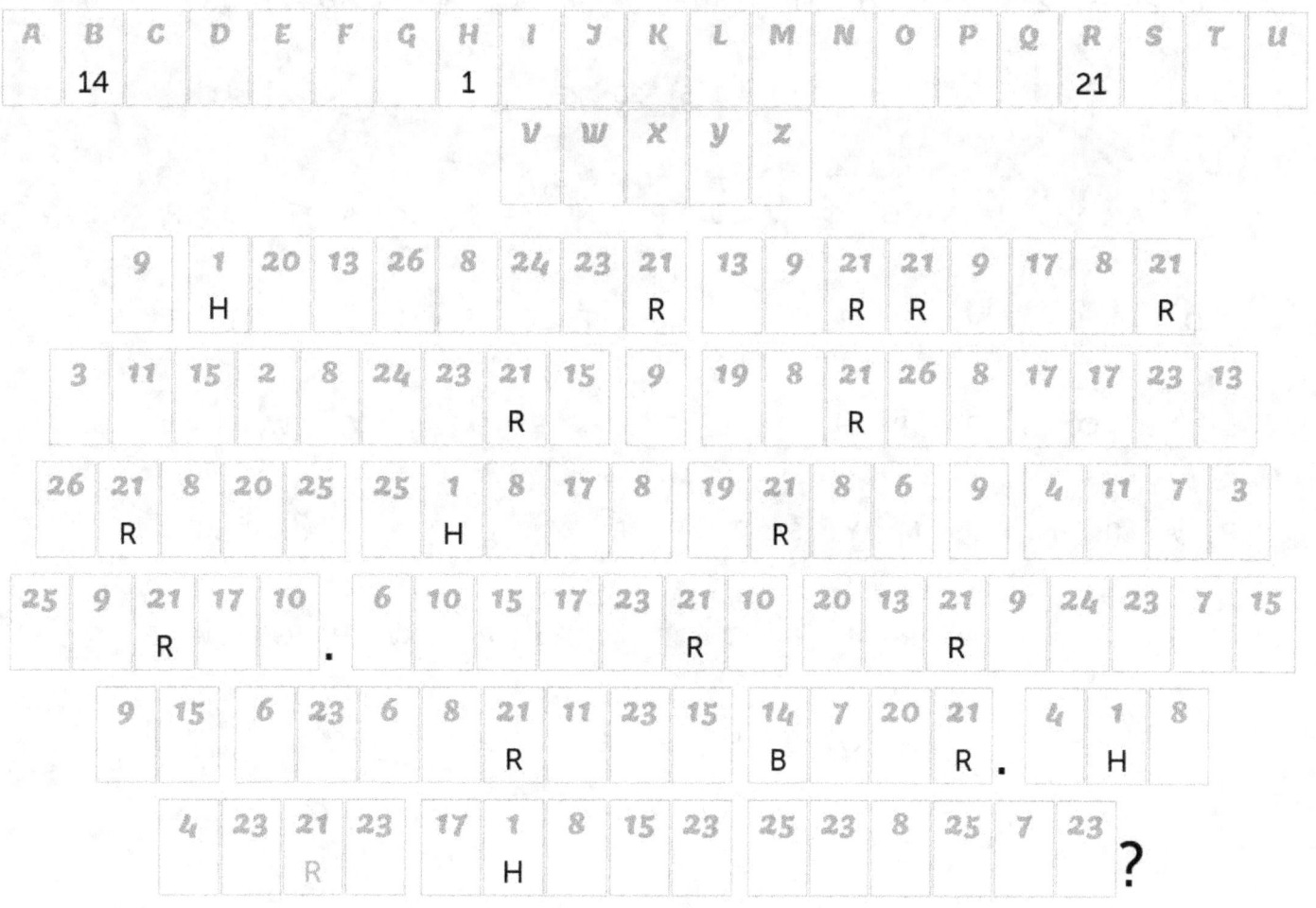

Answer.

A HUNGOVER NARRATOR DISCOVERS A FORGOTTEN GROUP PHOTO FROM A WILD PARTY. MYSTERY UNRAVELS AS MEMORIES BLUR. WHO WERE THOSE PEOPLE?

Quiz 25.   What is this secret message about?
Difficulty: * * * *

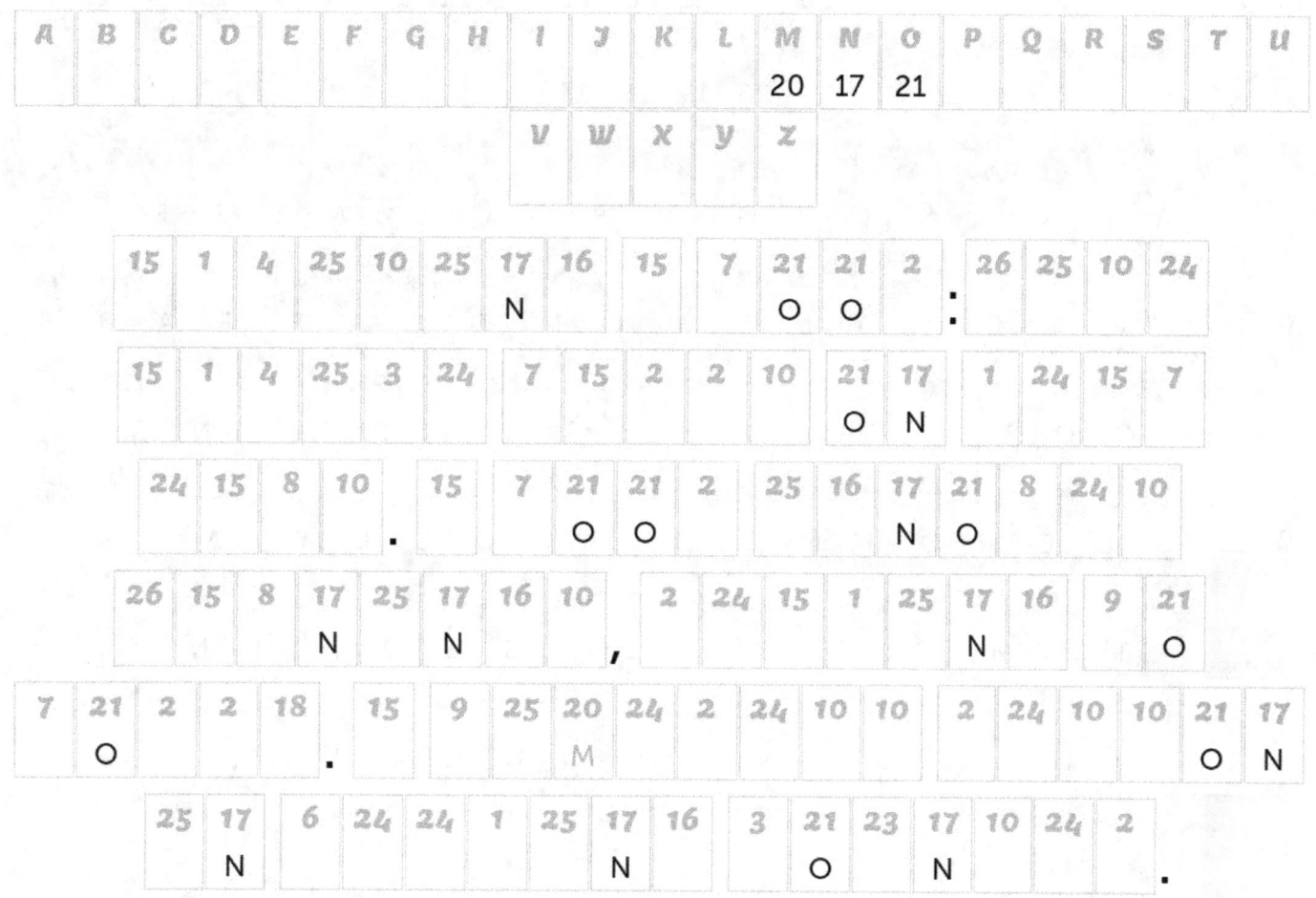

Answer.

A	B	C	D	E	F	G	H	I	J	K	L	M	N	O	P	Q	R	S	T	U	V
15	22	3	1	24	7	16	6	25	12	11	2	20	17	21	14	5	8	10	9	23	4

W	X	Y	Z
26	19	18	13

ADVISING A FOOL: WISE ADVICE FALLS ON DEAF EARS. A FOOL IGNORES WARNINGS, LEADING TO FOLLY. A TIMELESS LESSON IN HEEDING COUNSEL.

Quiz 26. What is this secret message about?
Difficulty: * * * *

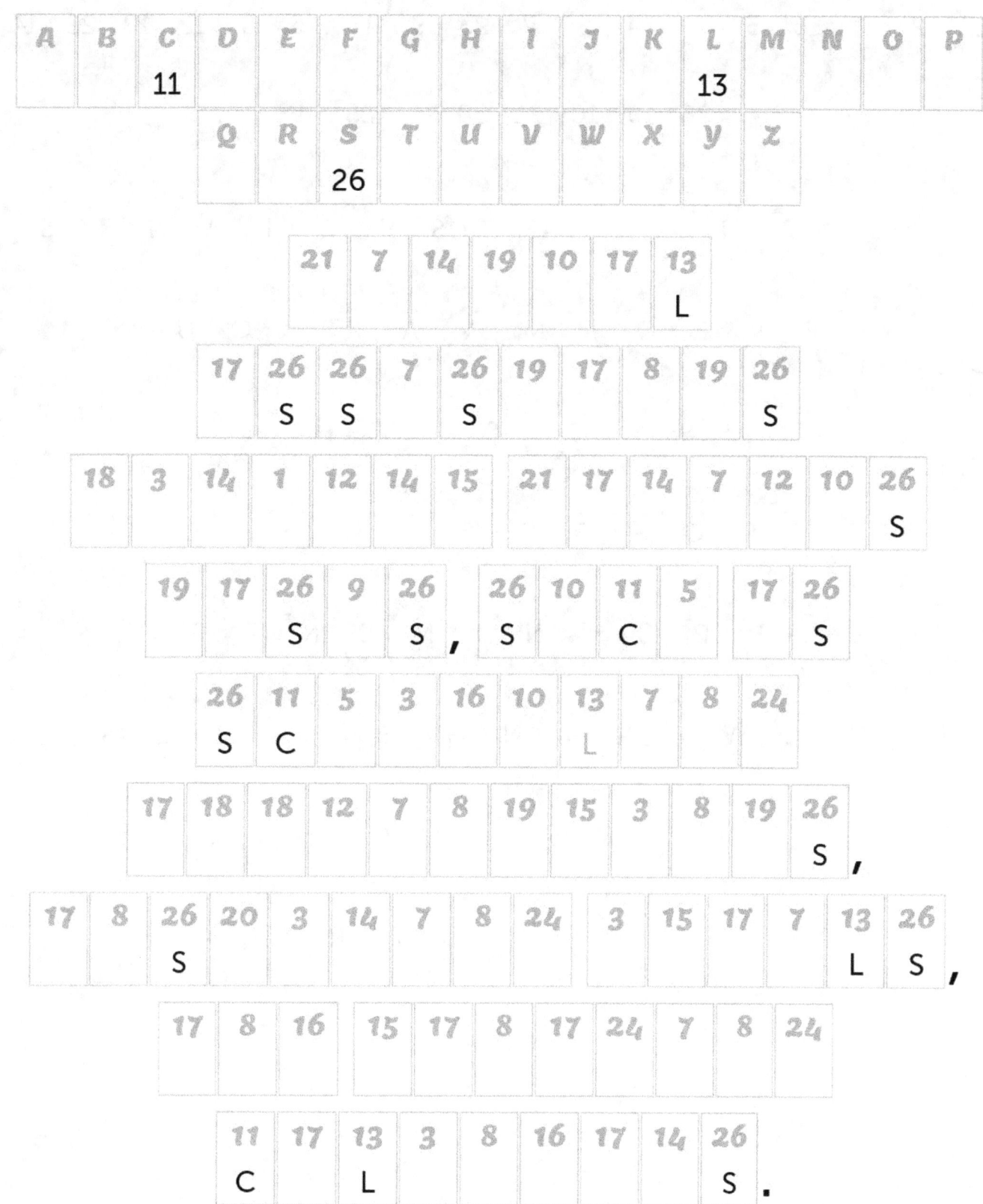

Answer.

VIRTUAL ASSISTANTS PERFORM VARIOUS TASKS, SUCH AS SCHEDULING APPOINTMENTS, ANSWERING EMAILS, AND MANAGING CALENDARS.

Quiz 27. What is this secret message about?
Difficulty: * * * *

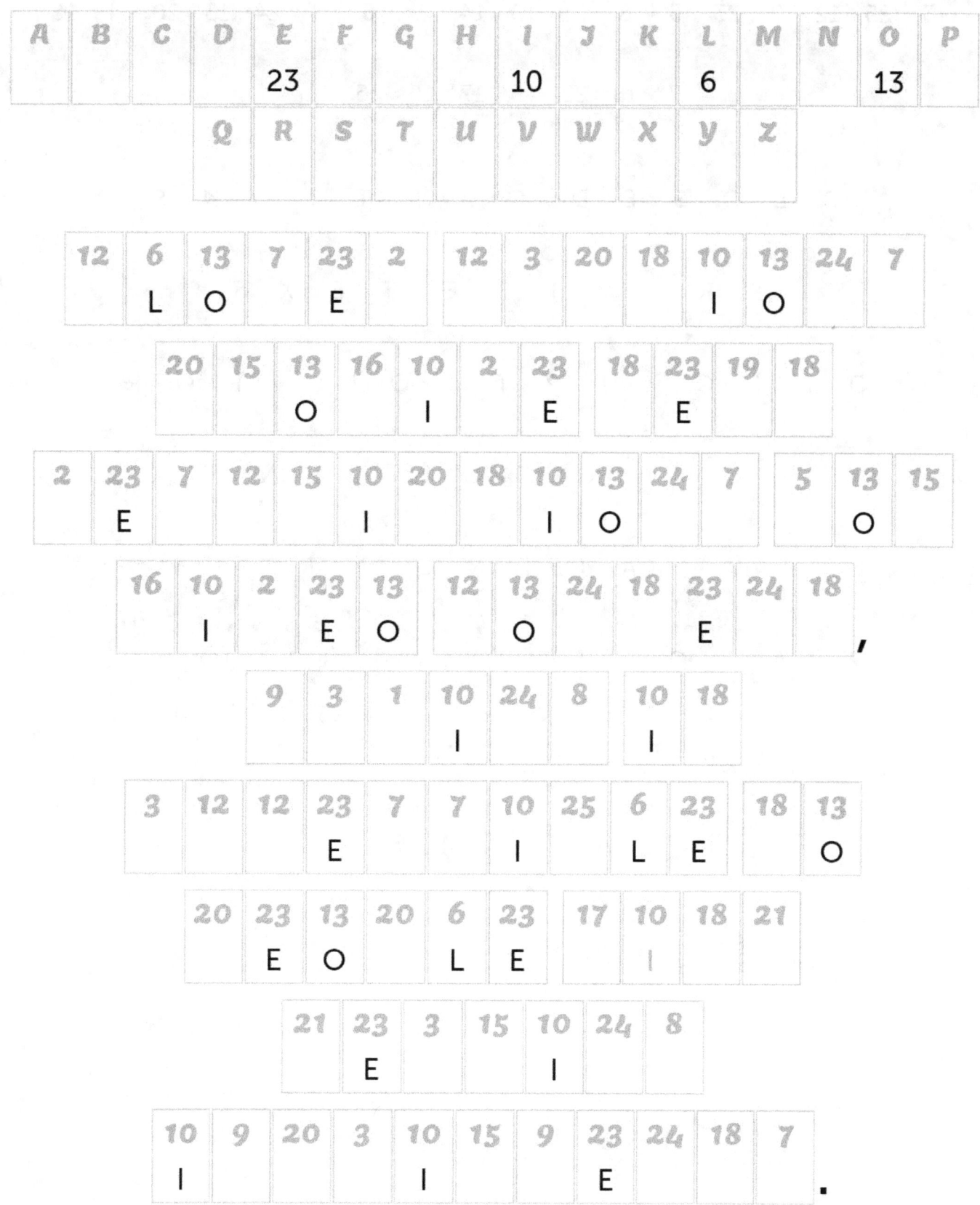

Answer.

A	B	C	D	E	F	G	H	I	J	K	L	M	N	O	P	Q	R
3	25	12	2	23	5	8	21	10	14	1	6	9	24	13	20	4	15

S	T	U	V	W	X	Y	Z
7	18	22	16	17	19	26	11

CLOSED CAPTIONS PROVIDE TEXT DESCRIPTIONS FOR VIDEO CONTENT, MAKING IT ACCESSIBLE TO PEOPLE WITH HEARING IMPAIRMENTS.

Quiz 28. What is this secret message about?
Difficulty: * * * *

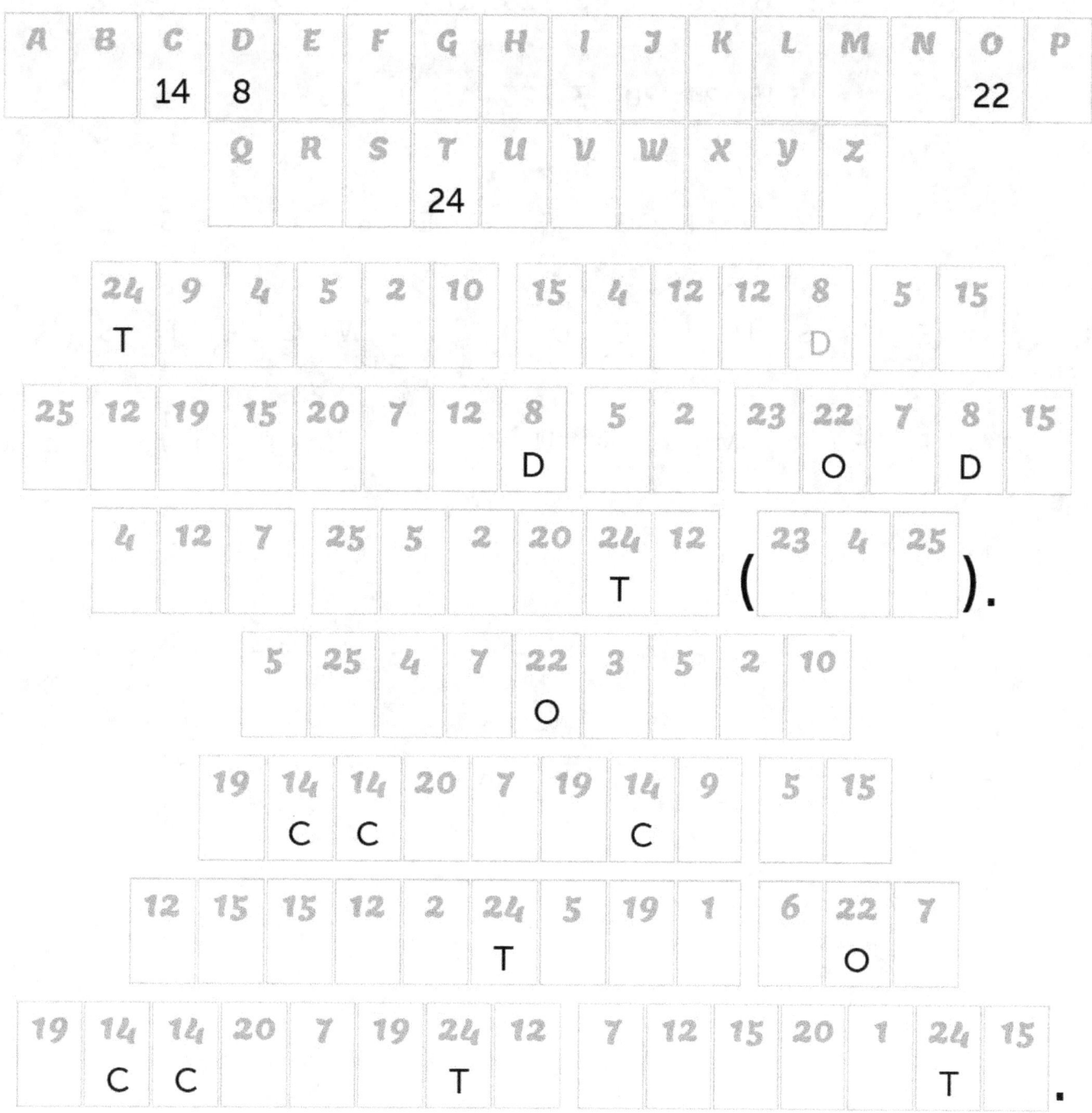

Answer.

TYPING SPEED IS MEASURED IN WORDS PER MINUTE (WPM). IMPROVING ACCURACY IS ESSENTIAL FOR ACCURATE RESULTS.

Quiz 29. What is this secret message about?
Difficulty: * * * *

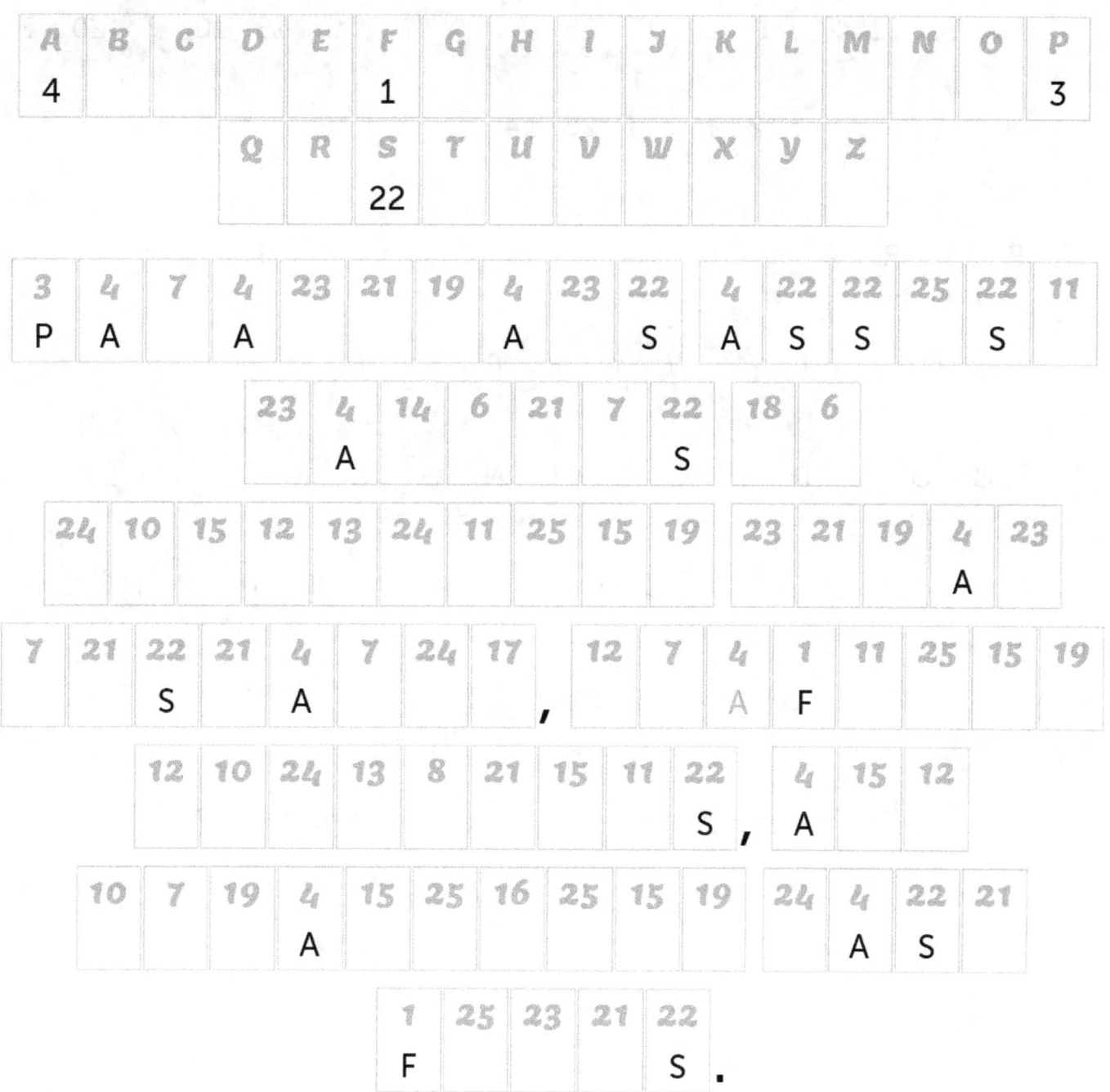

Answer.

A	B	C	D	E	F	G	H	I	J	K	L	M	N	O	P	Q	R
4	18	24	12	21	1	19	17	25	5	9	23	8	15	10	3	20	7

S	T	U	V	W	X	Y	Z
22	11	13	26	14	2	6	16

PARALEGALS ASSIST LAWYERS BY CONDUCTING LEGAL RESEARCH, DRAFTING DOCUMENTS, AND ORGANIZING CASE FILES.

Quiz 30. What is this secret message about?
Difficulty: * * * *

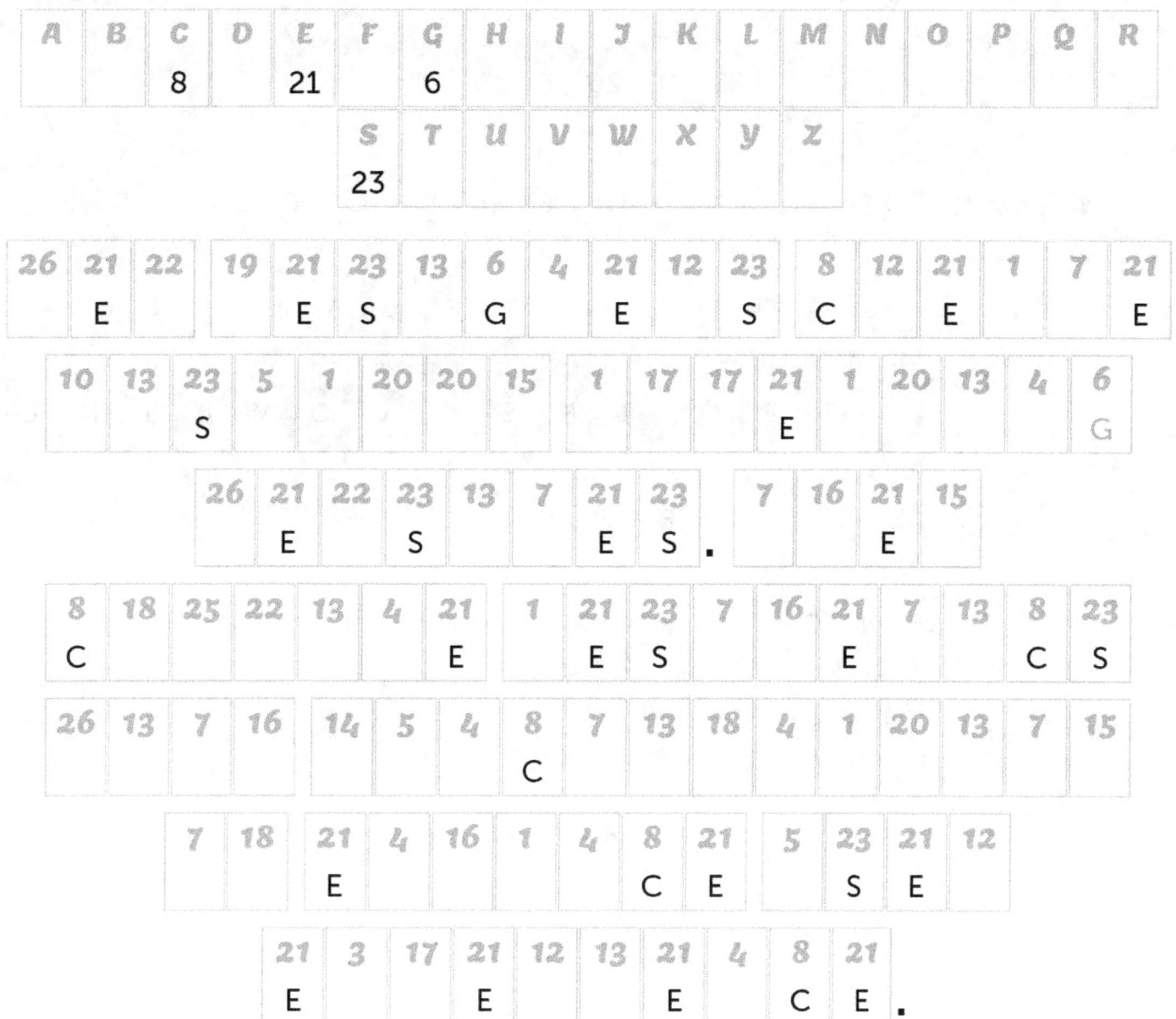

Answer.

A	B	C	D	E	F	G	H	I	J	K	L	M	N	O	P	Q	R	S	T
1	22	8	19	21	14	6	16	13	2	11	20	25	4	18	17	9	12	23	7

U	V	W	X	Y	Z
5	10	26	3	15	24

WEB DESIGNERS CREATE VISUALLY APPEALING WEBSITES. THEY COMBINE AESTHETICS WITH FUNCTIONALITY TO ENHANCE USER EXPERIENCE.

Quiz 31. What is this secret message about?
Difficulty: * * * * *

Answer.

A	B	C	D	E	F	G	H	I	J	K	L	M	N	O	P	Q	R
21	24	23	7	17	1	20	9	13	11	19	10	12	6	8	5	3	15

S	T	U	V	W	X	Y	Z
4	2	26	22	16	14	25	18

PROOFREADERS PLAY A CRUCIAL ROLE IN ENSURING WRITTEN CONTENT IS ERROR-FREE. THEIR JOB INVOLVES CHECKING GRAMMAR, SPELLING, AND PUNCTUATION.

Quiz 32.   What is this secret message about?
Difficulty: * * * * *

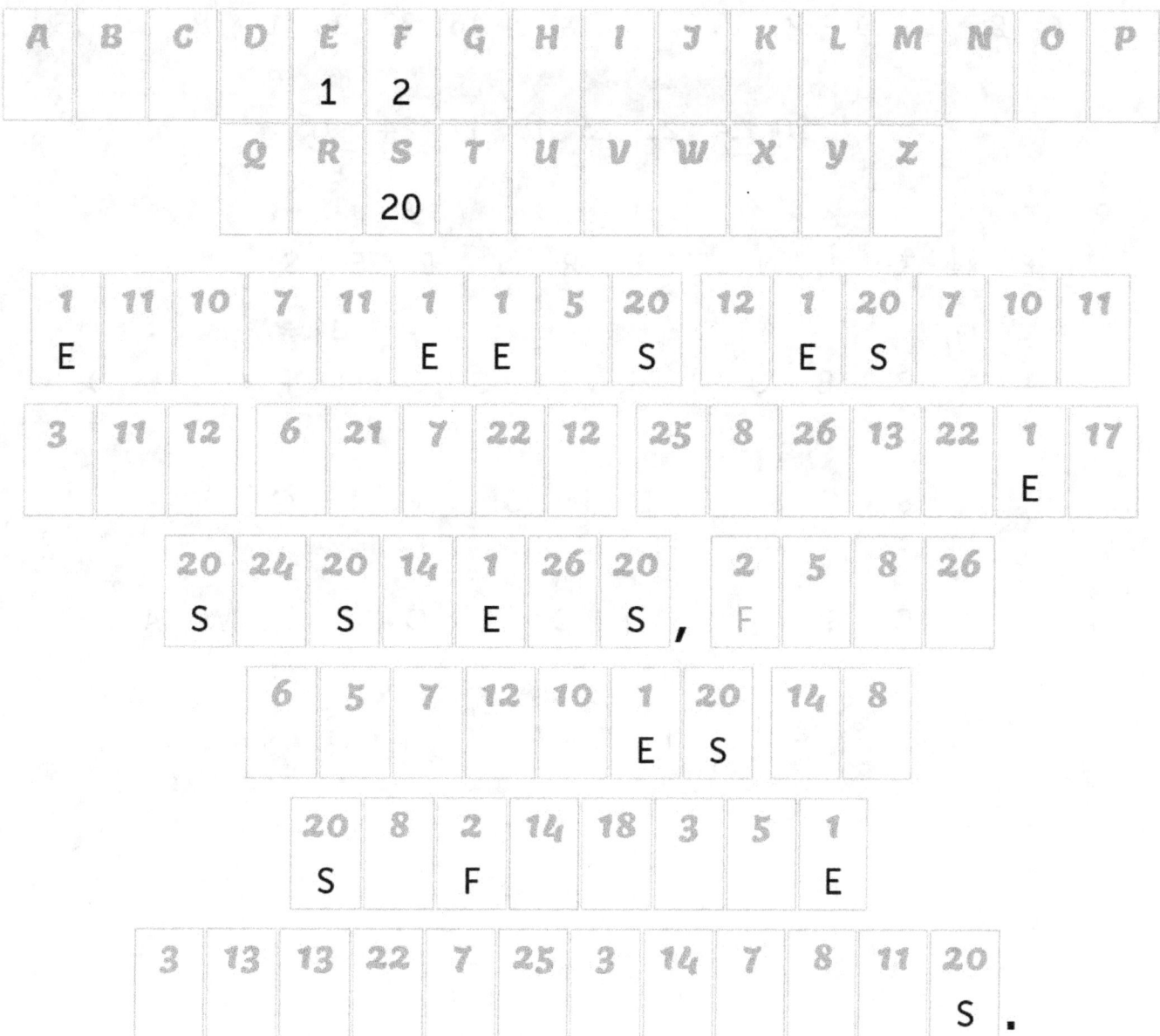

Answer.

Quiz 33.   What is this secret message about?
Difficulty: * * * * *

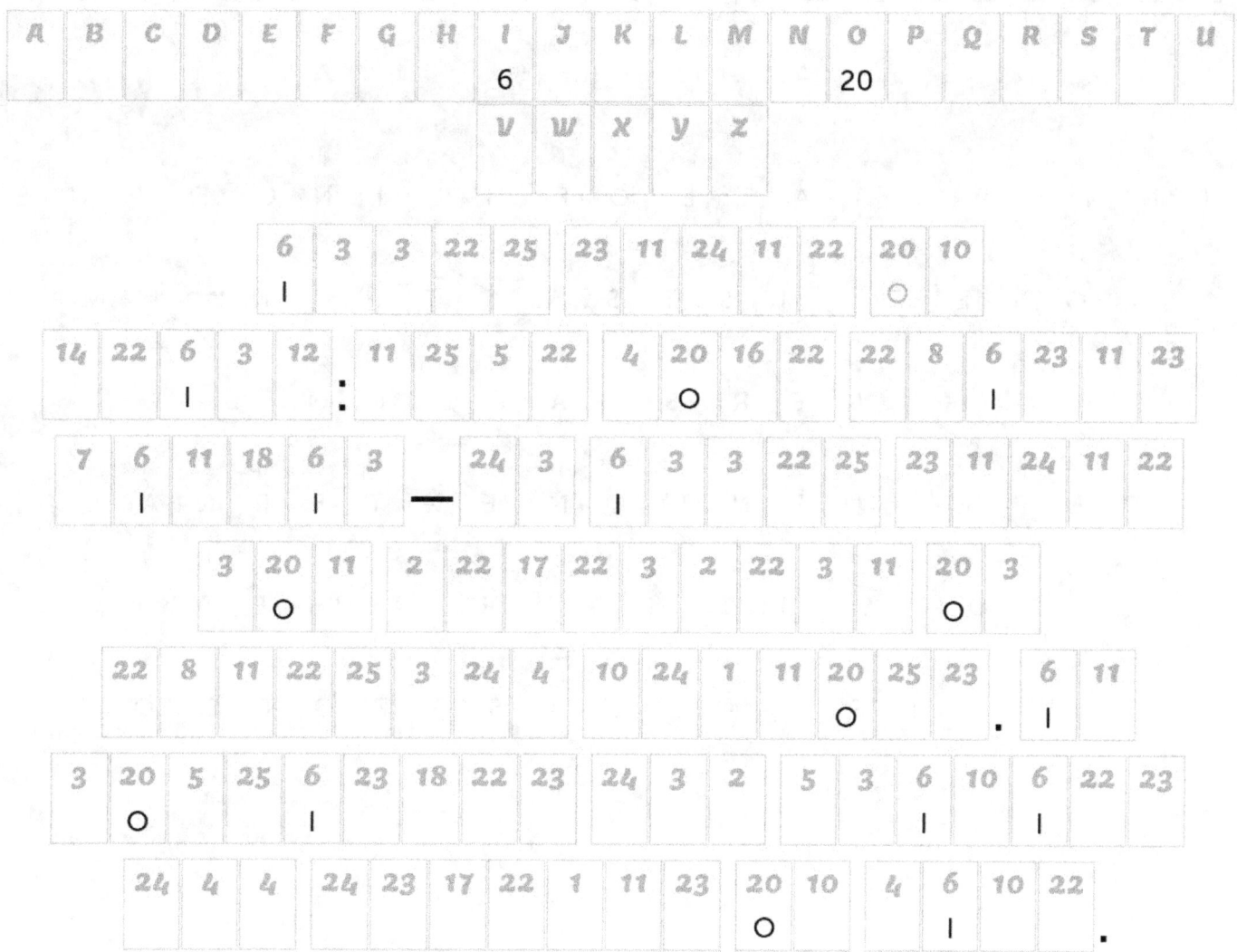

Answer.

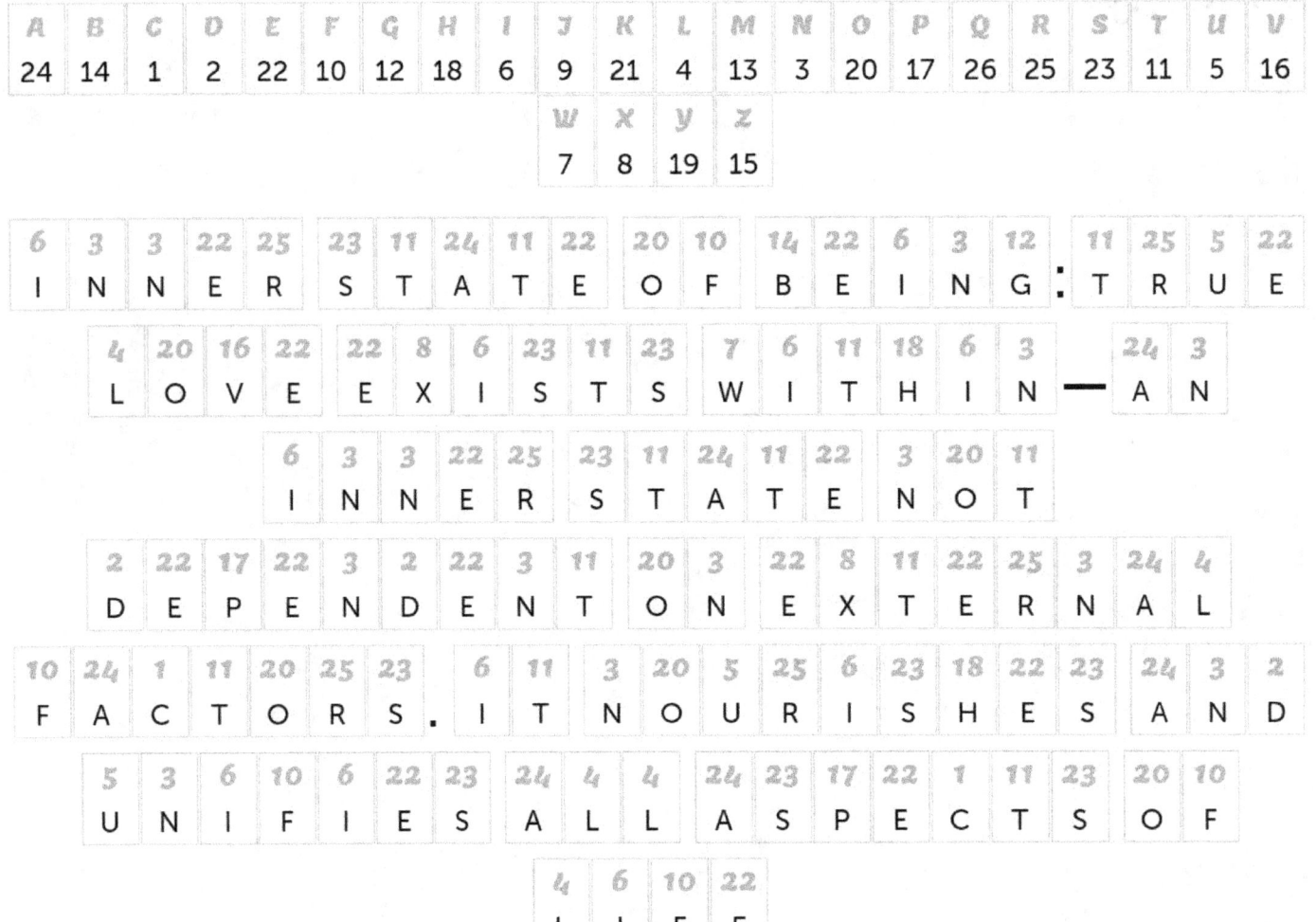

INNER STATE OF BEING: TRUE LOVE EXISTS WITHIN — AN INNER STATE NOT DEPENDENT ON EXTERNAL FACTORS. IT NOURISHES AND UNIFIES ALL ASPECTS OF LIFE.

Quiz 34. What is this secret message about?
Difficulty: * * * * *

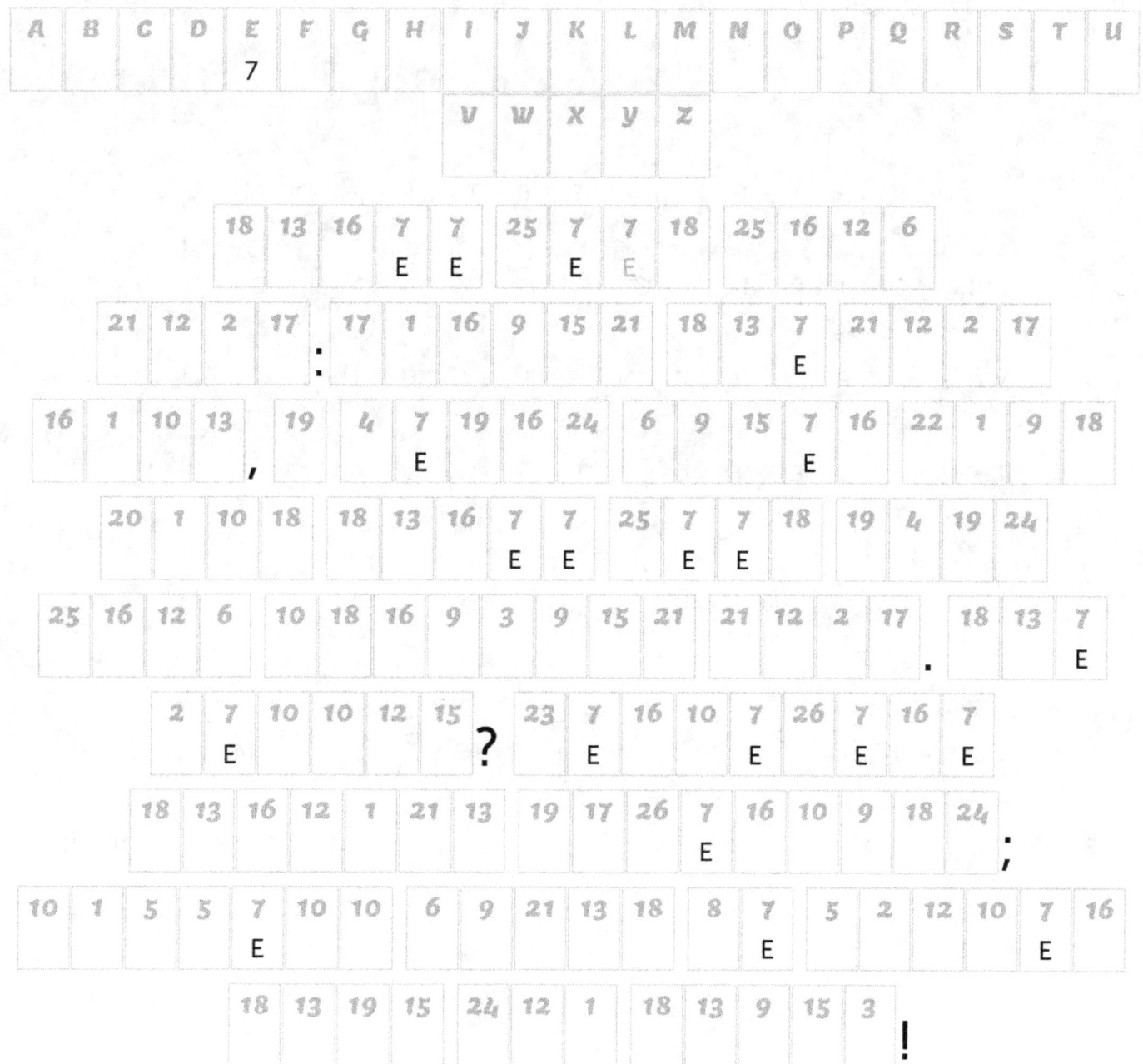

Answer.

A	B	C	D	E	F	G	H	I	J	K	L	M	N	O	P	Q	R	S	T	U	V
19	8	5	17	7	25	21	13	9	20	3	2	6	15	12	23	22	16	10	18	1	26

W	X	Y	Z
4	14	24	11

THREE FEET FROM GOLD: DURING THE GOLD RUSH, A WEARY MINER QUIT JUST THREE FEET AWAY FROM STRIKING GOLD. THE LESSON? PERSEVERE THROUGH ADVERSITY; SUCCESS MIGHT BE CLOSER THAN YOU THINK!

Quiz 35. What is this secret message about?
Difficulty: * * * * *

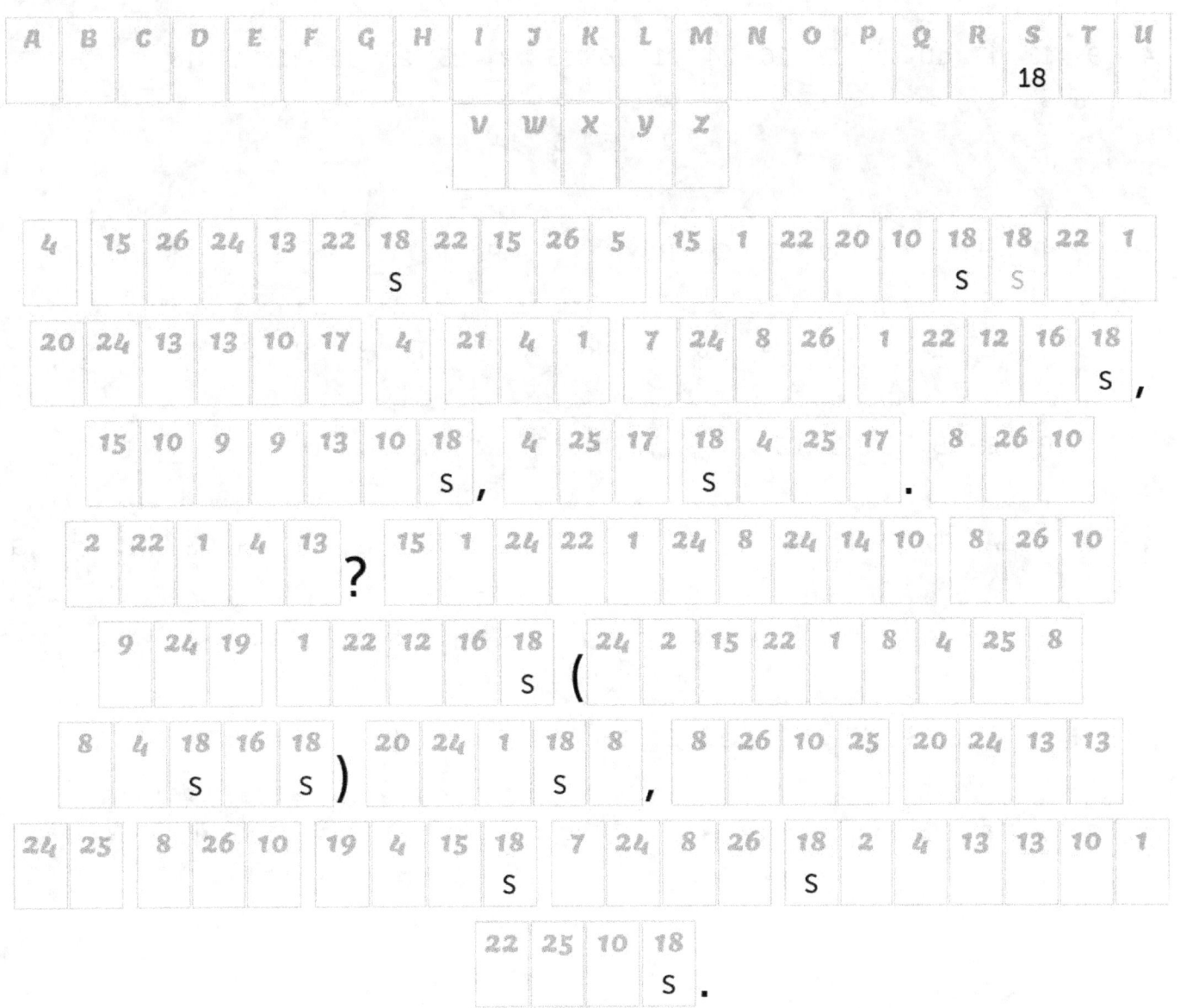

Answer.

A	B	C	D	E	F	G	H	I	J	K	L	M	N	O	P	Q	R	S	T	U	V
4	9	12	17	10	20	19	26	24	21	16	13	2	25	22	15	11	1	18	8	6	3

W	X	Y	Z
7	23	5	14

A PHILOSOPHY PROFESSOR FILLED A JAR WITH ROCKS, PEBBLES, AND SAND. THE MORAL? PRIORITIZE THE BIG ROCKS (IMPORTANT TASKS) FIRST, THEN FILL IN THE GAPS WITH SMALLER ONES.

Quiz 36.   What is this secret message about?
Difficulty: * * * * * *

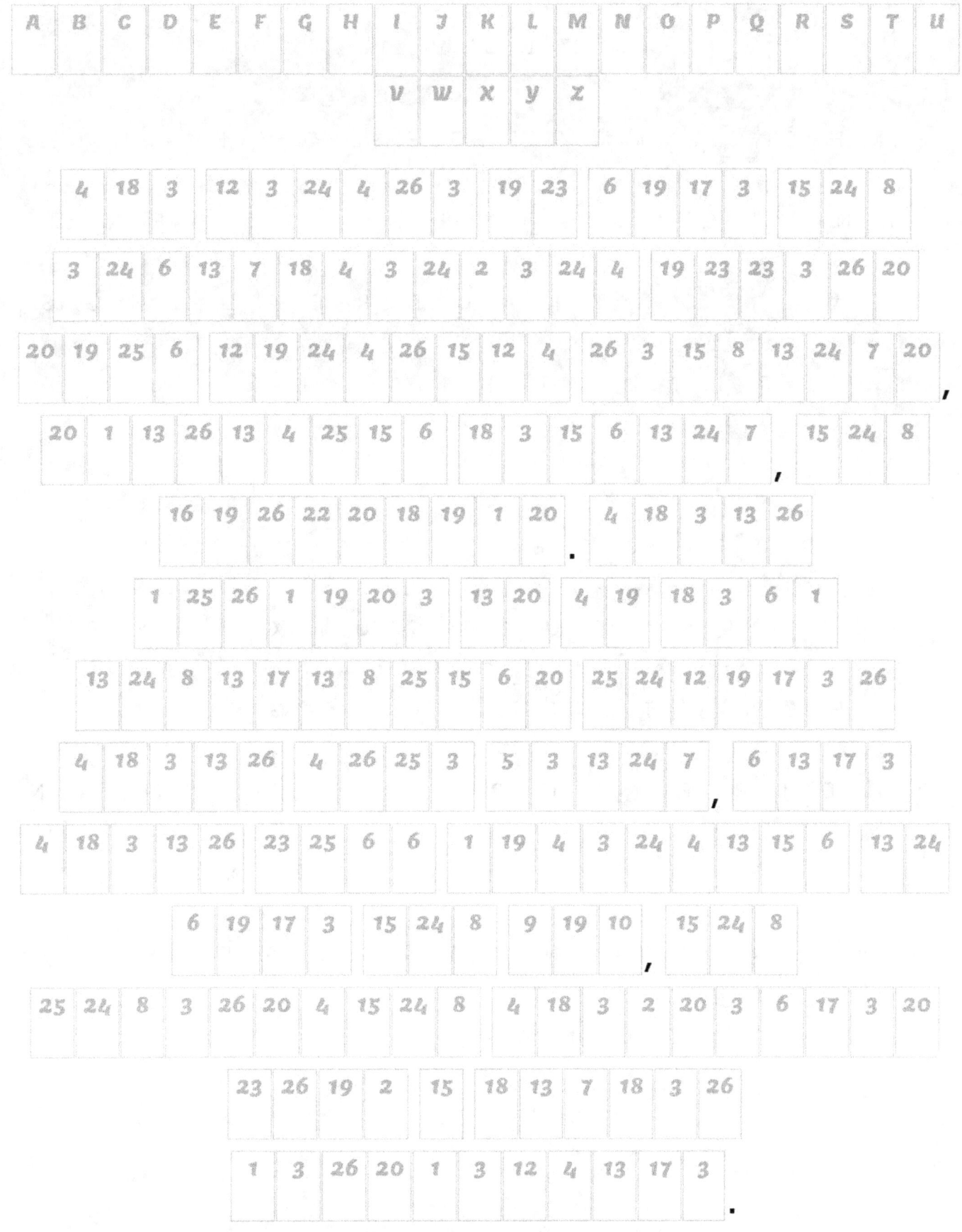

Answer.

A	B	C	D	E	F	G	H	I	J	K	L	M	N	O	P	Q	R	S	T	U
15	5	12	8	3	23	7	18	13	9	22	6	2	24	19	1	11	26	20	4	25

V	W	X	Y	Z
17	16	14	10	21

THE CENTRE OF LOVE AND ENLIGHTENMENT OFFERS SOUL CONTRACT READINGS, SPIRITUAL HEALING, AND WORKSHOPS. THEIR PURPOSE IS TO HELP INDIVIDUALS UNCOVER THEIR TRUE BEING, LIVE THEIR FULL POTENTIAL IN LOVE AND JOY, AND UNDERSTAND THEMSELVES FROM A HIGHER PERSPECTIVE.

Quiz 37. What is this secret message about?
Difficulty: * * * * * *

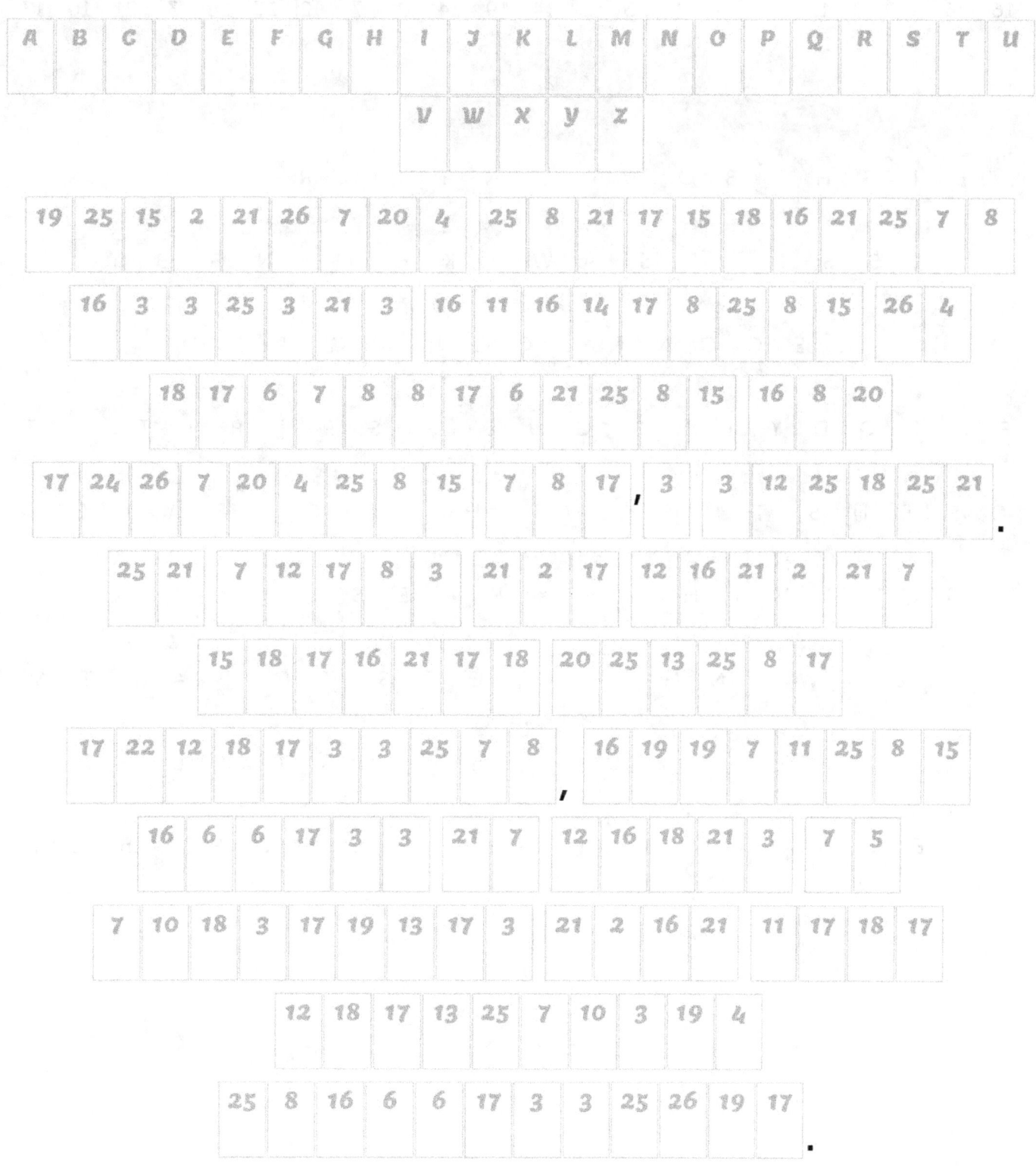

Answer.

A	B	C	D	E	F	G	H	I	J	K	L	M	N	O	P	Q	R	S	T	U	V
16	26	6	20	17	5	15	2	25	1	14	19	24	8	7	12	23	18	3	21	10	13

W	X	Y	Z
11	22	4	9

LIGHTBODY INTEGRATION ASSISTS AWAKENING BY RECONNECTING AND EMBODYING ONE'S SPIRIT. IT OPENS THE PATH TO GREATER DIVINE EXPRESSION, ALLOWING ACCESS TO PARTS OF OURSELVES THAT WERE PREVIOUSLY INACCESSIBLE.

Quiz 38. What is this secret message about?
Difficulty: * * * * * *

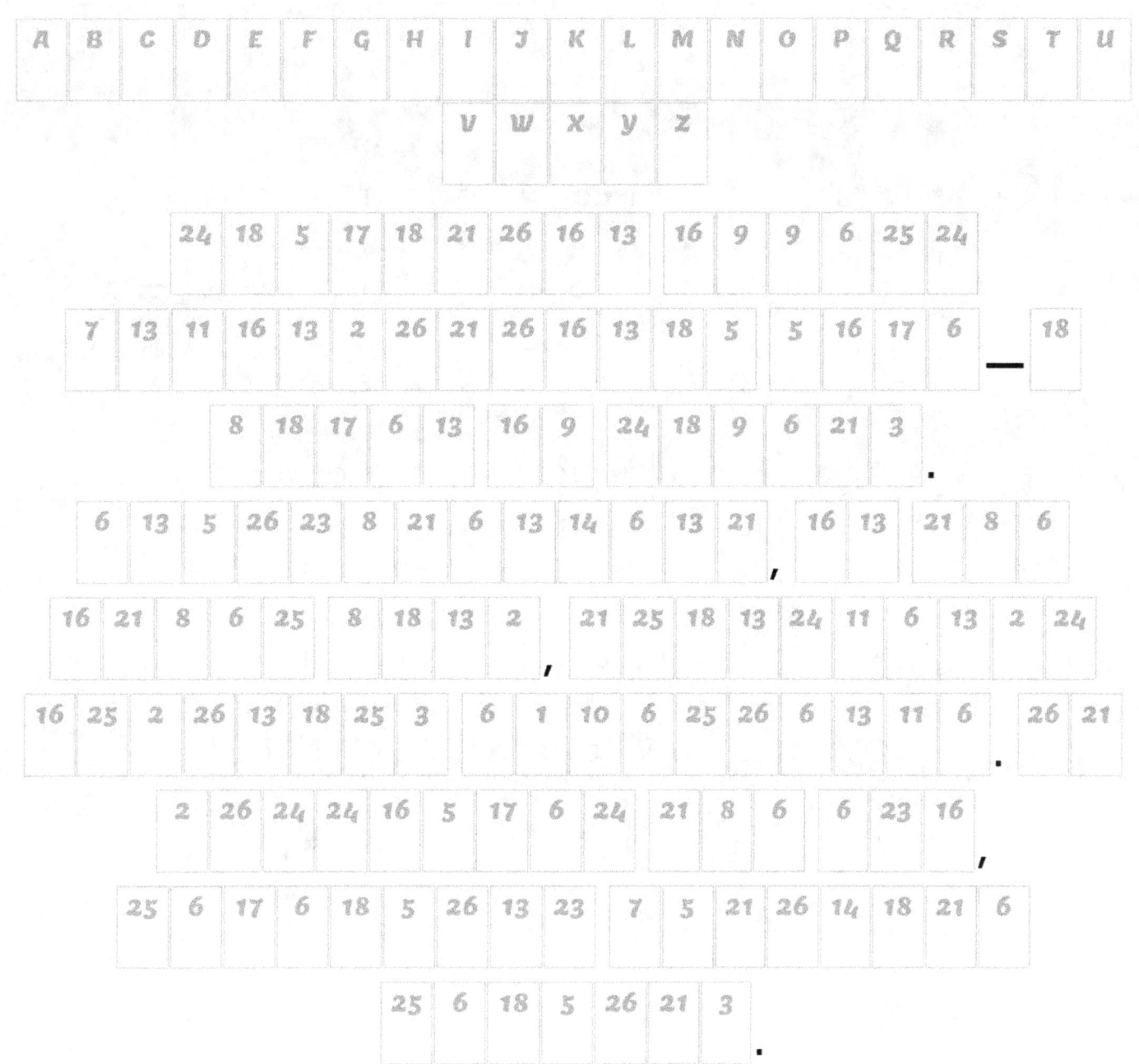

Answer.

A	B	C	D	E	F	G	H	I	J	K	L	M	N	O	P	Q	R	S	T	U	V
18	4	11	2	6	9	23	8	26	22	12	5	14	13	16	10	20	25	24	21	7	17

W	X	Y	Z
15	1	3	19

SALVATION OFFERS UNCONDITIONAL LOVE — A HAVEN OF SAFETY. ENLIGHTENMENT, ON THE OTHER HAND, TRANSCENDS ORDINARY EXPERIENCE. IT DISSOLVES THE EGO, REVEALING ULTIMATE REALITY.

Quiz 39. What is this secret message about?
Difficulty: * * * * * *

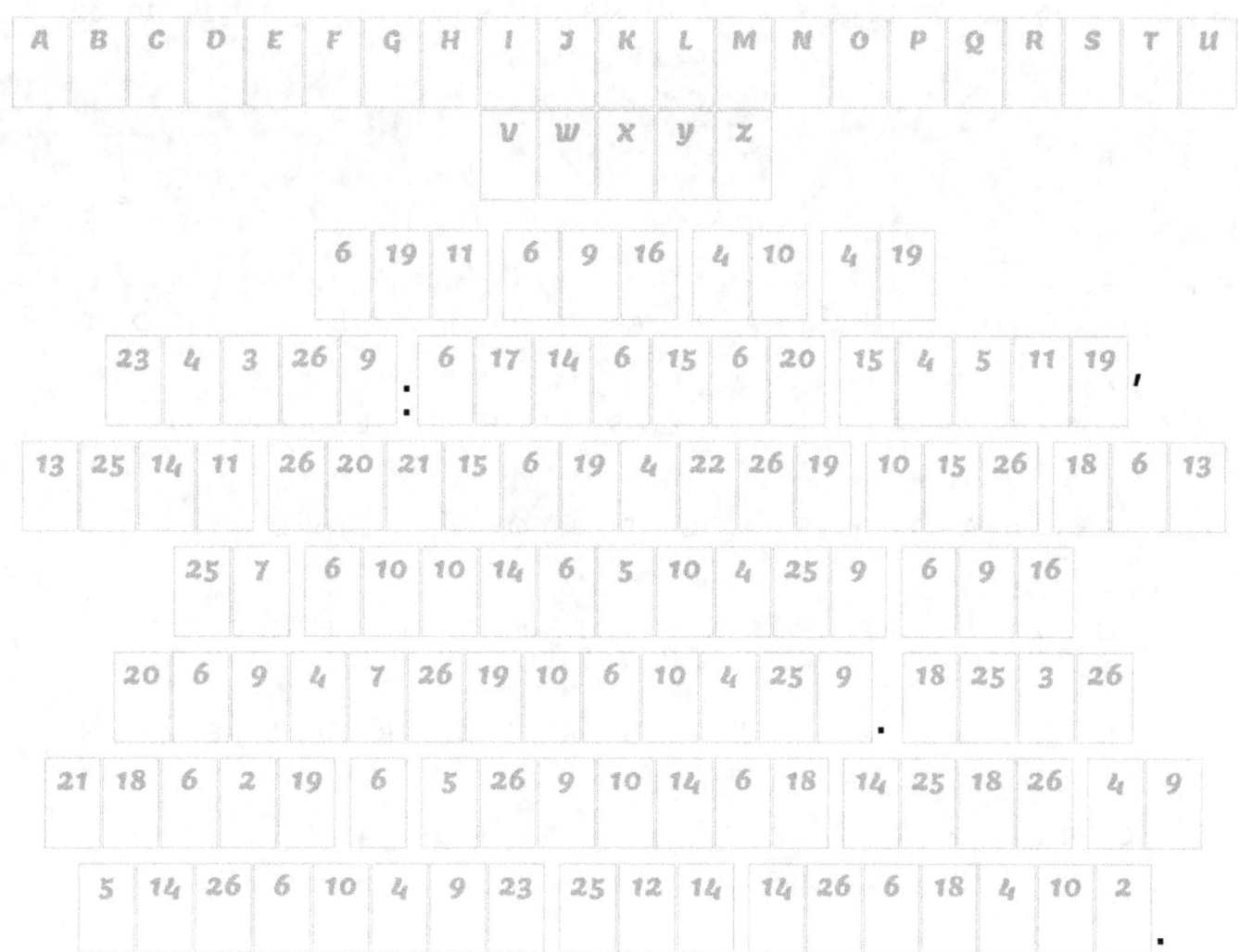

Answer.

ASK AND IT IS GIVEN: ABRAHAM HICKS' WORK EMPHASIZES THE LAW OF ATTRACTION AND MANIFESTATION. LOVE PLAYS A CENTRAL ROLE IN CREATING OUR REALITY.

Quiz 40.   What is this secret message about?
Difficulty: * * * * * *

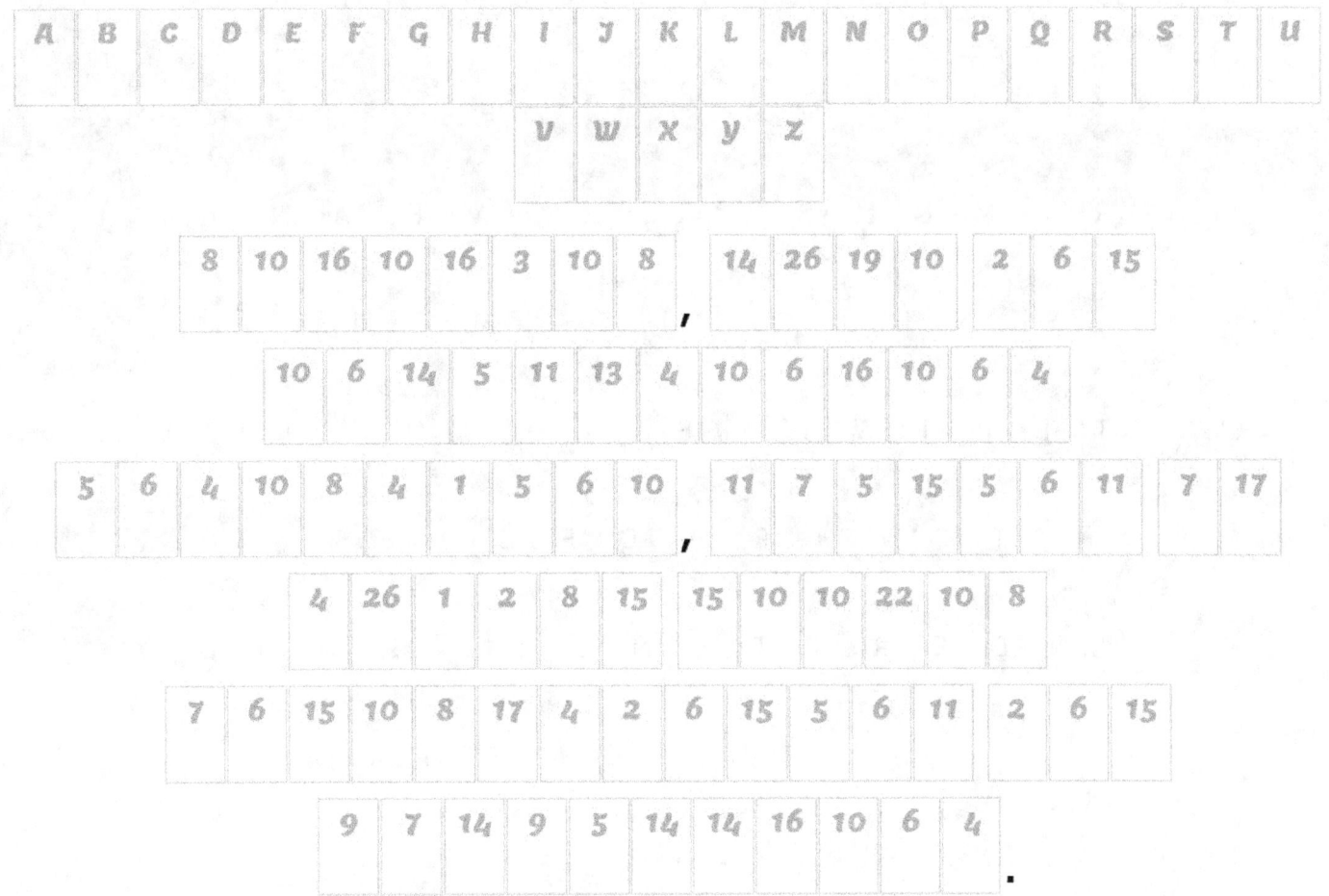

Answer.

REMEMBER, LOVE AND ENLIGHTENMENT INTERTWINE, GUIDING US TOWARD DEEPER UNDERSTANDING AND FULFILLMENT.

WELCOME TO

We feel honored that you've chosen **SCH Alyssa** to amplify your creative aspirations!

    Our ultimate goal is to sprinkle joy and enchantment into your day, eradicating any lingering stress.

    We treasure your satisfaction above all else, and we would be thrilled if you could take a moment to share your experience with us via a review on Amazon.

    In addition, please also take a moment to go to this link: https://forms.gle/aK42EtPGKxr97AyT6

and let us know more about how you feel. Don't hesitate; it would make our entire day! Your feedback help us polish our products to make them even better for you and future customers.

Copyright © 2023 by SCH Alyssa, Siu Chin Hung and
SCH Ali Good Fortune International Co

This book is safeguarded by copyright law, and all rights are reserved. Without the written permission of one of the authors, any reproduction or unauthorized usage of the material or artwork contained herein is strictly prohibited.

www.ingramcontent.com/pod-product-compliance
Lightning Source LLC
Chambersburg PA
CBHW082355220526
45470CB00008B/2753